T0212279

Principles of Secure Processor Architecture Design

Synthesis Lectures on Computer Architecture

Editor
Margaret Martonosi, *Princeton University*

Founding Editor Emeritus
Mark D. Hill, *University of Wisconsin, Madison*

Synthesis Lectures on Computer Architecture publishes 50- to 100-page publications on topics pertaining to the science and art of designing, analyzing, selecting and interconnecting hardware components to create computers that meet functional, performance and cost goals. The scope will largely follow the purview of premier computer architecture conferences, such as ISCA, HPCA, MICRO, and ASPLOS.

Principles of Secure Processor Architecture Design

Jakub Szefer

ISBN: 978-3-031-00632-6   paperback
ISBN: 978-3-031-01760-5   ebook
ISBN: 978-3-031-00057-7   hardcover

DOI 10.1007/978-3-031-01760-5

A Publication in the Springer series
SYNTHESIS LECTURES ON ADVANCES IN AUTOMOTIVE TECHNOLOGY

Lecture #45
Series Editor: Margaret Martonosi, *Princeton University*
Founding Editor Emeritus: Mark D. Hill, *University of Wisconsin, Madison*
Series ISSN
Print 1935-3235   Electronic 1935-3243

Principles of Secure Processor Architecture Design

Jakub Szefer

Yale University

SYNTHESIS LECTURES ON COMPUTER ARCHITECTURE #45

ABSTRACT

With growing interest in computer security and the protection of the code and data which execute on commodity computers, the amount of hardware security features in today's processors has increased significantly over the recent years. No longer of just academic interest, security features inside processors have been embraced by industry as well, with a number of commercial secure processor architectures available today. This book aims to give readers insights into the principles behind the design of academic and commercial secure processor architectures. Secure processor architecture research is concerned with exploring and designing hardware features inside computer processors, features which can help protect confidentiality and integrity of the code and data executing on the processor. Unlike traditional processor architecture research that focuses on performance, efficiency, and energy as the first-order design objectives, secure processor architecture design has security as the first-order design objective (while still keeping the others as important design aspects that need to be considered).

This book aims to present the different challenges of secure processor architecture design to graduate students interested in research on architecture and hardware security and computer architects working in industry interested in adding security features to their designs. It aims to educate readers about how the different challenges have been solved in the past and what are the best practices, i.e., the principles, for design of new secure processor architectures. Based on the careful review of past work by many computer architects and security researchers, readers also will come to know the five basic principles needed for secure processor architecture design. The book also presents existing research challenges and potential new research directions. Finally, this book presents numerous design suggestions, as well as discusses pitfalls and fallacies that designers should avoid.

KEYWORDS

secure processor design, processor architecture, computer security, trustworthy computing, computer hardware security

Dla ukochanej Injoong i najwspanialszej Adusi.

Contents

Preface

Recent years have brought increased interest in hardware security features that can be added to computer processors to protect sensitive code and data. It has been realized that new hardware security features can be used to provide, for example, means of authentication or protection of confidentiality and integrity. The hardware offers a very high level of immutability, helping to ensure that it is difficult to change the hardware security protections (unlike with software-only protections). Hardware cannot be as easily bypassed or subverted as software, as it is the ultimate lowest layer of a computer system. Finally, dedicated hardware for providing security protections can potentially offer energy efficiency and minimal impact on system performance.

Yet, adding security features in hardware has many challenges. Defining what has to be secured, and how, is often a subjective choice based on qualitative arguments—unlike the quantitative choices that computer architects are used to making. Moreover, once made, the hardware cannot be easily changed, which necessitates careful design of the security features in the first place. The secure architecture design process also requires foresight to include features and algorithms that will be suitable for many years to come. Perhaps the biggest challenges are the attacks and various information leaks that the system should protect against. Not only random errors or faults need to be considered, but the system also needs to defend against "smart" attackers who can intelligently manipulate inputs or probe the hardware to try to maximize their chances of subverting the computer system's protections.

This book assumes readers may be at the level of a first- or second-year graduate student in computer architecture. The book is also suitable for more senior students or for practicing computer architects who are interested in starting work on the design of secure processor architectures. The book provides a chapter on security topics such as encryption, hashing, confidentiality, and integrity, to name a few—consequently a background in computer security is not required. It is the hope that this book will get computer architects excited about security and help them work on secure processor architectures.

The chapters of this book are based on research ideas developed by the author and also ideas gleaned from papers that a variety of researchers have presented in conferences such as ISCA, ASPLOS, HPCA, CCS, S&P, and Usenix Security. Information is also included about recent commercial architectures, such as Intel SGX, ARM TrustZone, and AMD memory encryption technologies. The book, however, is not meant as a manual or tutorial about any one specific security architecture. Rather, it uses past academic and industry research to derive and present the principles behind design of such secure processor architectures.

This book is divided into ten chapters. Chapter 1 focuses on motivating the need for secure processor architectures and gives an overview of the book's organization. Chapter 2 covers basics of computer security needed for understanding secure processor architecture designs. It can be considered an optional chapter for those already familiar with major computer security topics. Chapter 3 discusses main features of secure processor architectures, such as extending processors with new privilege levels, or breaking the traditional linear hierarchy of the privilege levels. Chapter 4 focuses on the Trusted Execution Environments which are created by the hardware and software Trusted Computing Base, and discusses various protections that secure architectures can offer to the Trusted Execution Environments. Chapter 5 introduces the Root of Trust from which most of the security features of a secure processor architecture are derived. Chapter 6 is an in-depth discussion of protections that secure architectures use to protect main memory, usually DRAM. Chapter 7 overviews security features that target designs with many processors or many processor cores. Chapter 8 gives extended review of side channel threats, processor features that contribute to existence of side channels, and ideas for eliminating various side channels. Chapter 9 is an optional chapter, which can be considered a mini survey of work on security verification of processor architectures and hardware. Chapter 10 concludes the book by presenting the five principles for secure processor architecture design, along with research challenges and future trends in secure processor designs.

After finishing this book, readers should be familiar with the five design principles for secure processor architecture design, numerous design suggestions, as well as become educated about pitfalls and fallacies that they should avoid when working on secure processor designs. Most importantly, security at the processor and hardware level is a crucial aspect of today's computers, and this book aims to educate and excite readers about this research area and its possibilities.

Jakub Szefer
October 2018

Acknowledgments

The ideas and principles derived in this book are based not only on my own research, but also on research and ideas explored over many years by numerous researchers and gleaned from their academic papers presented in top architecture and security conferences. I would like to especially acknowledge my former Ph.D. adviser, Prof. Ruby B. Lee, and others with whom I learned about, and worked on, secure processor architectures. The principles and ideas presented here reflect the hard work of many researchers and of the broader computer architecture and security communities.

I would like to thank Prof. Margaret Martonosi, the editor of the Synthesis Lectures on Computer Architecture series, and Michael B. Morgan, President and CEO of Morgan & Claypool Publishers, for their support and deadline extensions. I hope this book is a valuable addition to the series, and it was made much better through their input and encouragement. In addition, this book was improved thanks to the feedback and reviews from Margaret Martonosi, Caroline Trippel, and Chris Fletcher. Further, I would like to thank Dr. C.L. Tondo, Christine Kiilerich, and the copyeditors for helping bring this book to reality.

Work on this book was made possible in part through generous support from the National Science Foundation, through grants number 1716541, 1524680, and an NSF CAREER award number 1651945, and through support by Semiconductor Research Corporation (SRC). It was further made possible through support from Yale University.

Special thanks to my current Ph.D. students: Wenjie Xiong, Wen Wang, Shuwen Deng, and Shanquan Tian. It is a pleasure to work with them on secure processor architectures, hardware security, and other topics related to improving computer hardware security; our work forces me to constantly learn new ideas and push the boundaries on these exciting research topics.

I would like to thank my parents, Ewa and Krzysztof, for their constant encouragement, especially during my years in graduate school, and now in my academic career. Their unwavering love and support can always be counted on.

Most importantly, I would like to thank my amazing wife, Injoong, for all she does. Without her, my research, work, and this book would not be possible. She is the most loving wife and my best friend. And last, but not least, many hugs and kisses to our baby daughter, Adriana, for being the cutest and smartest baby ever! Every day is a surprise and she brings nothing but joy to me.

Jakub Szefer
October 2018

CHAPTER 1

Introduction

This chapter provides motivation for research and work on secure processor architectures. It also provides an outline of the organization of this book and, in particular, highlights the core and the optional chapters.

1.1 NEED FOR SECURE PROCESSOR ARCHITECTURES

Secure processor architectures by design provide extra hardware features which enhance commodity processors with new security capabilities. The new security features may be purely in hardware, or they may be implemented in both hardware and software. The former are the so-called hardware security architectures, and the latter are the so-called hardware-software architectures. Unlike hardware security modules or dedicated security accelerators (see Section 3.6.5), secure processor architectures are mainly designed as extensions of commodity processors, and are based on architectures such as such x86 or RISC. The increased need for implementing security features in processor architectures has been driven in recent years by three factors.

- **Software Complexity and Bugs**: Increased complexity and size of the software code running on commodity processors, especially the operating system or the hypervisor code, makes it impractical, or even impossible, to provide security solely based in software—more and more lines of software code lead to increased number of software bugs and potential exploits. New features (e.g., new protection levels or new hardware features for creating trusted software executing environments) are needed to provide an execution environment wherein a small, trusted code can execute separated from the rest of the untrusted code.

- **Side-Channel Attacks**: Computation is today often done in settings such as in cloud computing where many different users share the same physical hardware. Co-residency of potential victims and attackers on same hardware can allow the attackers to learn sensitive information through shared hardware and the side channels. Timing-based side channels, and also power, RF, or EM-based ones, exploit known side effects of the behavior of commodity processors when different types of computations are performed. Only modifications at the architecture and hardware levels can mitigate different types of side channels and the resulting side-channel attacks.

- **Physical Attacks**: Attacks including physical probing of memory busses or even memory chips have necessitated protections against not just software, but hardware or physical attacks, especially as cloud computing has increased in popularity, where users no longer have physical control over the hardware on which their code runs, new mechanisms are needed to protect the code and data against physical attacks. Likewise, embedded devices or Internet-of-Things devices are prone to physical attacks as users may not have physical control over the devices at all times.

The first motivating factor for the need for secure processor architectures listed above is the constantly increasing code size of the software, and consequent number of bugs and exploits. According to some estimates there can be as many as 20 bugs per 1000 lines of software code [69]. Even if it is an overestimate, an operating system or hypervisor with millions of lines of code will have thousands of bugs. The operating system is typically in charge of managing applications, and as long as the operating system is running correctly, applications are protected from one another. Today, however, the code base of an average operating system has grown to tens of millions of lines of code and it is no longer possible for operating system to be bug free—the operating system is no longer trustworthy to protect the applications. Similarly, after introduction of the new hypervisor privilege level, hypervisors were supposed to be small and trustworthy to isolate different operating systems form one another. Yet today, hypervisors are more like operating systems with millions of lines of code. The bloated code base makes the operating systems and hypervisors less trustworthy. To secure applications (from each other and from the untrustworthy operating system) or to secure virtual machines (VMs) (from each other and from the untrustworthy hypervisor), a variety of secure processor architectures, and design ideas, have been proposed that leverage changes in the architecture. New privilege levels or mechanisms for separation of code and data can help protect trusted code from the rest of the code running on the computer system.[1]

The second motivating factor for the need for secure processor architectures is side-channel attacks. Side-channel attacks that leverage timing, power, RF, or EM changes or emanations as processors execute different programs and instructions. The side-channel attacks have been known and explored for a long time. Defending against such types of attacks, however, has gained new urgency after Spectre [120] and Meltdown [138] attacks were publicized (while this book was being written), which partly leverage cache timing side channels, for example.

The third factor driving the need for secure processor architectures is the physical attacks. Lack of physical control over the computer system where the code is executing means that users can no longer be sure that there is physical security and that nobody is tampering with their system. For example, in cloud computing, users rent servers or VMS that are located in far away data centers where rogue employees or hosting company compelled by the government can probe the server hardware while it is running. In another scenario, embedded devices, such as in now popular Internet-of-Things computing paradigm, are spread out in variety of locations where the owner may not be able to ensure that they are physically secure.

---

[1]Hardware designs can suffer form bugs in the hardware description code, just as there are bugs in software. Chapter 9 gives information about approaches for security verification of processor architectures and designs.

1.2   BOOK ORGANIZATION

Organization of this book's chapters is shown in Figure 1.1 below. The overarching design goal of secure architectures is to protect integrity and confidentiality of user applications, operating system, hypervisor, or other software components, depending on the threat model and assumptions of the particular architecture, and prevent software or hardware attacks (again, within limits of the particular threat model). To help students and practitioners learn about, and create, such solutions and protections, the remainder of the book is divided into a number of chapters focusing on major topic areas, culminating in the last chapter that presents the principles for secure processor architecture design.

The remaining chapters of this book can be divided into three groups. Chapter 2 covers basics of computer security needed for understanding secure processor architecture designs. It can be considered an optional chapter for those already familiar with topics such as: confidentiality, integrity, availability, symmetric and public-key cryptography, secure hashing, hash trees, etc. Chapters 3–8 and also Chapter 10 are the core of this book. Chapter 9 is again an optional chapter, which can be considered a mini survey of work on security verification of processor architectures and hardware.

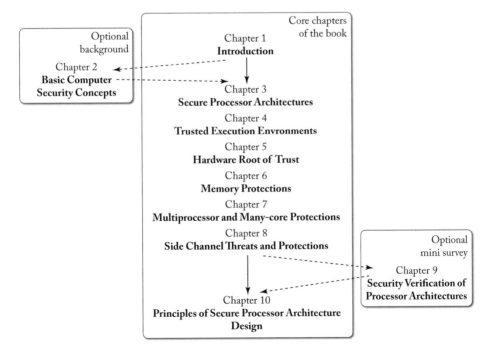

Figure 1.1: Organization of this book's chapters.

CHAPTER 2

Basic Computer Security Concepts

This chapter covers basics of computer security needed for understanding secure processor architecture designs. It begins with a discussion of the trusted computing base, security threats to a system, and discussion of threat models. A brief overview of information leaks and side channels is also included in the chapter as it will be needed to understand the side-channel attack protections that secure processor architectures needs. This chapter then dives into security concepts of confidentiality, integrity, and availability. It also explains basics of symmetric-key cryptography, public-key cryptography, and secure hashes. A short section on importance of good sources of randomness is included. The chapter closes with a short overview of physically unclonable functions (PUFs) and their applications.

2.1   TRUSTED COMPUTING BASE

In a secure processor architecture, the Trusted Computing Base (TCB) is formed by the hardware components and the software components that work together and provide some security guarantees, as specified by the architecture. There is a distinction that can be made between hardware secure architectures where the hardware is protecting the software but all security mechanisms are solely in hardware, or hardware-software secure, architectures where hardware works with some software (usually privileged software such as the operating system or the hypervisor) to protect other software. The latter can provide more flexibility, but typically increases the size of the TCB. Larger TCB is usually considered less desirable as more lines of software code (and likewise lines of hardware description code) are assumed to be correlated with more potential bugs [69] and consequently security vulnerabilities.

A typical computer system can be broken down into a number of distinct hardware components, e.g., processor cores, processor caches, interconnect, Dynamic Random Access Memory (DRAM), etc. These hardware components interact with each other as well as with the different software components as the system executes. The hardware components which are dependent upon to provide security form the hardware TCB. Meanwhile, the software components (if any) which are dependent upon to provide security form the software TCB. These hardware and software components that work together and provide some security guarantees are assumed to be trusted, and together form the whole TCB.

Correct operation of the system depends on the correctness of the TCB. The goal of secure processor architectures is then to ensure that the trusted hardware components can work together with the trusted software components to provide the desired security properties and protections for the software (e.g., trusted software modules or enclaves, whole user applications, or even containers or VMs) running on the system.

In addition to the trusted hardware and software components, there are the untrusted parts of the system. The untrusted parts (hardware or software) need not be overtly malicious, but are simply not trusted for the correct operation of the system. During the design of a secure processor architecture, the trusted computing base has to be constructed such that, regardless of the actions taken by the untrusted parts, the security properties of the system will be maintained. Effectively, each untrusted entity can be a potential attacker that tries to break the security of the system—the designer has to then consider all possible attacks by all the untrusted entities, unless a specific threat is explicitly not protected against, as specified in the threat model (discussed in Section 2.2.6). Beyond the untrusted entities which are part of the system, there are external attackers which should also be considered, i.e., physical attacks on the computer system.

It should be emphasized that trusted parts are ones on which the correct operation of the system explicitly depends on. If something happens to a trusted part (e.g., a software function is altered or a hardware module is modified) then the protections of the whole system can no longer be assumed. The trusted part, however, could be malicious to begin with or buggy due to poor design. A trusted software or a trusted hardware part thus may or may not be trustworthy.

Trustworthiness is a qualitative designation indicating whether the entity will behave as expected, is free of bugs and vulnerabilities, and is not malicious. The designation of an entity as trustworthy is separate from the designation of the entity as trusted. A secure processor architecture is designed with explicit assumption about which hardware or software entities need to be trusted, i.e., these entities from the TCB. During the design and implementation of these entities, it needs to be ensured that they are indeed trustworthy. Techniques such as formal security verification [52] should be applied to make sure the design is correct. However, even beyond design, bugs, or malicious modifications can be introduced during manufacturing time, e.g., hardware trojans can be added [213].

Architecture designers typically focus on ensuring that the protocols, interactions, interfaces, and use of encryption and hashing among the trusted components are such that there can be no attack. Once there is confidence that the system cannot be attacked due to a logical design in the flaw, the focus can move to the implementation details. Implementation details[1] focus on issues such as malicious foundries [231] or supply chain security [178]. Even when the lifetime of a processor ends, issues of trustworthiness can continue (e.g., make sure the system permanently destroys the encryption keys).

---

[1]It should be noted that side channels are related to both the design process, e.g., timing side channels due to the way the cache is designed are independent of the physical implementation details of the cache, and to the implementation process, e.g., different types of transistors or logic gates may create thermal or EM side channels. Side channels are discussed in Section 2.2.4.

2.1.1   KERCKHOFFS'S PRINCIPLE: AVOID SECURITY THROUGH OBSCURITY

Kerckhoffs' principle is well known and was first created in the context of cryptographic systems [167]. The principle can be paraphrased as stating that operation of the TCB of a security system should be publicly known and should have no secrets other then the cryptographic keys. Thus, even if an attacker knows everything about the operation of the TCB, he or she still cannot break the system unless he or she knows the cryptographic keys.

Many failed security systems practice the opposite of this principle, which is security through obscurity. Security through obscurity can be paraphrased as attempting to secure the system by making the operation of the TCB secret and hoping that any potential attackers are not able to reverse engineer the system and break it. Designers should not underestimate the cleverness of attackers and they should never practice security through obscurity. Security through obscurity has led to many real attacks, such as on metro cards [44], or could potentially lead to attacks on computer processors, as with researchers recently breaking into Intel's Management Engine [60] that runs secretive code which is in charge of the computer platform.

2.2   SECURITY THREATS TO A SYSTEM

The processor, the whole hardware of the computer system, and the software executing on it can be vulnerable to a number of security threats. The attackers can exploit the attack surface to mount different types of attacks, and these need to be protected against.

2.2.1   THE ATTACK SURFACE

The attack surface is the combination of all the attack vectors that can be used against a system. Individual attack vectors are different ways that an attacker can try to break system's security. Figure 2.1 shows different types of attack vectors (left side of the figure), along with potential parts of the system which can be targets of the attacks (bottom of the figure). Attack vectors could be through hardware or software, and in both cases could be due to so-called external attackers (attacker is not executing code on the target computer nor physically near the computer system they are attacking) or so-called internal attackers (the attacker is running code on the system he or she is trying to attack, or has physical access to the system). Hardware attacks could come from untrusted hardware (not in the TCB), or external physical attacks (e.g., physical probing of the memory or data buses). Software attacks could come from untrusted software (not in the TCB) or the software that is supposed to be protected (but due to bugs or malicious behavior tries to attack the system on which it is running). The targets of the attacks, shown at the bottom of Figure 2.1, can be hardware which is in the TCB, software which is in the TCB, or the software that is supposed to be protected. It does not make sense to try to attack the untrusted hardware or the untrusted software as by definition it does not provide any security mechanisms nor hold any sensitive data.

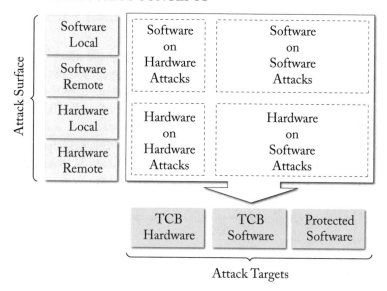

Figure 2.1: The potential attack surface of a secure processor. The terms external and internal refer to whether the potential attack is from within the system or from outside. The targets of the attacks can be either the TCB or the software that is being protected (by the TCB).

There are numerous examples of the different classes of attacks. Software-on-software attacks could be untrusted operating system attacking software that is being protected. A software-on-hardware attack could be untrusted software using cache side-channel attacks to learn secret information from a processor cache's operation. A hardware-on-software attack could be an untrusted memory controller trying to extract information from DRAM memory. A hardware-on-hardware attack could be untrusted peripheral trying to disable memory encryption engine. Note that components inside the software or hardware TCB are never assumed to be sources of attacks, as by definition, they are trusted with ensuring protections for the system.

2.2.2 PASSIVE AND ACTIVE ATTACKS

Passive attacks are the types of attacks where the attacker only observes the operation of the system. He or she does not actively trigger any inputs nor otherwise interact with the system. Passive attackers are most often called an eavesdroppers. They listen in on communication (internal or external to the system) to try to deduce the secret information or code. Side-channel attacks are types of passive attacks where attackers gather timing measurements, or power, EM, thermal, or other emanations from the system to try to learn what it is doing or to learn the secrets (more on side channels is mentioned shortly in Section 2.2.4).

Active attacks involve the attacker trying to modify code or data of the system. An attacker may try to write into some memory location (e.g., to change encryption key stored there) or execute a series of instructions (e.g., to heat up the chip and cause it to fail). Physical attacks on processor chips such as fault injection are also active attacks as the attacker actively manipulates the (physical) state of the system.

The attacks can be further divided into: snooping, spoofing, splicing, replay, and disturbance attacks. Snooping attacks are passive attacks where the attacker simply tries to observe or read some information. Spoofing attacks are active attacks that, for example, involve injecting new memory commands to try to read or modify data in memory. Splicing attacks are also active attacks, which, for example, involve combining portions of legitimate memory commands (observed before through a snooping attack) into new memory commands (to read or modify data). Typically, splicing involves combining requests from one command (e.g., a memory read from, or a write to, a specific address) with authentication information from another command. Replay attacks are another example of active attacks which involve, for example, re-sending previously observed memory command again in the future. Disturbance attacks are the last type of active attack and they include attacks such as Denial of Service (DoS) on memory bus [153], using repeated memory accesses to age circuits [110], repeated memory access to trigger Rowhammer bug [117], etc.

2.2.3   MAN-IN-THE-MIDDLE ATTACKS

Man-in-the-middle attacks are attacks on communication (internal external to the system). As the name implies, man-in-the-middle attempts to intercept communication between two trusted components, and the goal is for neither entity to recognize that someone is intercepting the communication. Man-in-the-middle can be passive (receive data, read it, and forward to original destination without any changes) or active (modify data after receiving it or inject some new data).

2.2.4   SIDE AND COVERT CHANNELS AND ATTACKS

A covert channel [67] is a communication channel that was not intended or designed to transfer information between a sender and a receiver. Covert channels typically leverage unusual methods for communication of information, never intended by the system's designers. These channels can include use of timing, power, thermal emanations, electro-magnetic (EM) emanations, acoustic emanations, and possibly others. With the exception of timing channels, most channels require some physical proximity and sensors to detect the transmitted information, e.g., use of EM probe to sense EM emanations. Meanwhile, many timing-based channels are very powerful as they do not require physical access, only that sender and receiver run some code on the same system. Some covert channels can be prevented at the design time if they are known to exist, e.g., through proper isolation of processes or by partitioning caches, but many go unnoticed until the system is deployed.

Covert channels are important when considering intentional information exfiltration where one program manipulates the state of the system according to some protocol and another observers the changes to read the "messages" that are sent to it. Covert channels are a concern because even when there is explicit isolation, e.g., each program runs in its own address space and cannot directly read and write another program's memory, a covert channel may allow the isolation mechanisms to be bypassed.

A side channel [128] is similar to a covert channel, but the sender does not intend to communicate information to the receiver, rather the sending (i.e., leaking) of information is a side effect of the implementation and the way the computer hardware or software is used. Side channels can use same means as covert channels, e.g., timing, to transmit information. Typically, covert channels and attacks are analyzed or presented first as both sender and receiver is under control of the potential attacker and it is easier to create a covert channel. Next, side channels are usually explored as they are more difficult to create since the victim (i.e., sender) is not under control of the attacker. Some side channels, like covert channels, can be prevented at the design time, e.g., through constant-time software implementations or by randomizing caches, but many also go unnoticed until the system is deployed.

Side channels are important when considering unintentional information leaks. In a side channel, there is usually a victim process that uses a computer system and the way the system is used can be observed by an attacker process. Processor-based side and covert channels can be generally categorized as timing-based, access-based, or trace-based channels. Timing-based channels rely on the timing of various operations to leak information, e.g., [4, 21, 122]. For example, one process performs many memory accesses so that memory accesses of another process are slowed down. Access-based channels rely on accessing some information directly, e.g., [83, 161, 165, 174, 253]. For example, one process probes the cache state by timing its own memory accesses. Trace-based channels rely on measuring exact execution of a program, e.g., [1]. For example, attacker obtains sequence of memory accesses and whether they are cache hits or misses based on the power measurements. Besides processor-based side and covert channels, others exist, such as through power [121], EM radiation [6], thermal [106], etc.

Both covert and side channels can be created as a result of the logical design of the system, or due to the implementation of the system. Timing channels are often due to design, e.g., caches have different timing for a cache hit vs. cache miss, thus observation of the timing of memory access can reveal some information about state of the system. Timing channels can often be fixed or at least mitigated by changing the design, e.g., disable caches or use partitioned or randomized caches. Implementation-related channels can be, for example, EM channels, where the logical design does not leak any information (e.g., no timing cache channel exists), but EM radiation can reveal some information (e.g., memory address used to perform a memory access). Implementation-related channels, similar to hardware trojans, cannot often be easily defended at the architecture level. Meanwhile, timing channels or other due to the logical design of the system can be defended by changing the system design.

2.2.5   INFORMATION FLOWS AND ATTACK BANDWIDTHS

The information flow refers to the transfer of information between different entities. Information flow can be explicit, e.g., $a = b$ where data or information in b is moved to a, or it can be implicit, e.g., $b = 0$; *if*(a) *then* $b = 1$, where the value of b reflects whether a is true, but there was never a direct assignment, or copying of data, from a to b. Typically when discussing information flow there is a low-security entity that interacts with a high-security entity. A system could have a desired property such as "there is no information flow between the components x and y" or "component x's file z is never accessible by component y." Information flow can happen through data or through timing information.

If there is a possible information flow between the system and an attacker, then there is a potential attack vector. This has been formalized in security verification by number of tools which can try to check whether there are information flows between different entities in the system [52]. Side and covert channels can also be expressed in terms of information flow between an attacker and a victim.

When there is information flow, one needs to also consider bandwidth and the probabilistic nature of the information flow. Bandwidth simply means how much data can be transferred in a unit of time. Naturally, higher bandwidth information flows are more dangerous from security perspective. Although, in many cases to goal of an attacker is to exfiltrate a cryptographic key, e.g., 128–4096 bits, and the actual bandwidth does not need to be very high to get useful information (i.e., the key). The other aspect is the probabilistic nature of the information flow. As an example, in a cache side-channel attack, there is interaction between attacker process and victim process, who both affect the state of the cache, which leads to the side-channel attack. However, the information flow depends on the behavior of the attacker and the victim. If victim and attacker always access mutually exclusive cache sets, there is no information leak; if they access or contend for the same cache set, there can be information leak. Thus, there is not always an actual information flow, even it it is possible to have one.

Whether there is an information flow between a potential attacker and victim can be analyzed using non-interference. Non-interference is a property that typically refers to how untrusted entities interact with trusted entities. The interference between these entities can be analyzed to check that the untrusted entity is not able to observe any differences in its own behavior, or the behavior of the system, in response to, or as a byproduct of, a trusted entity processing sensitive or non-sensitive inputs. The trusted entity, however, may observe differences in the behavior of system or the untrusted entity. Non-interference essentially means that information about the operation of the secure entity does not leak to the insecure entity.

2.2.6   THE THREAT MODEL

A threat model is a concise specification of the threats that a given secure processor architecture protects against. It is unlikely that an architecture can be designed to protect against all possible hardware and software attacks. Also, it may be economically infeasible to try to provide protections against certain, unlikely attacks. At the very least, a threat model should specify the assumptions and threats that the architecture considers. It should specify the following.

1. TCB: the set of trusted hardware and software components.

2. Security properties: properties that the TCB aims to guarantee.

3. Attacker assumptions: capabilities of potential attackers.

4. Potential vulnerabilities: attackers and attack vectors on the trusted computing base, including untrusted entities and also any external attackers which will be defended against.

As part of the threat model, it is beneficial to also list any groups of potential attackers, attackers' capabilities, or attack vectors which are out-of-the-scope of the protections the trusted computing base of the secure architecture aims to ensure. This can clarify what is not being protected by the design.

Different secure processor architectures will provide protections under assumptions of different possible attacks, thus direct comparison of the architectures is often difficult. For example, some may protect from physical attacks on DRAM memory but not side channels, but others may protect against timing side channels, but not against physical attacks.

When there is an attack, since each architecture is designed for a specific threat model, it is important to distinguish whether the attack is indeed within the scope of the threat model. Sometimes the problem with the architecture is actually with the threat model. For example, recent Intel SGX architecture does not aim to protect against side-channel attacks due to caches [104], while there are many researchers presenting such attacks against SGX [75]. Architects need to consider needs and expectations of the users and make sure the threat model matches what is assumed by the users.

2.2.7   THREATS TO HARDWARE AFTER THE DESIGN PHASE

Secure processor design focuses on minimizing the TCB, and protecting against a variety of attacks. All the protections depend on the trustworthiness of the TCB, and at the design time the designer should attempt to verify the TCB [52]. When the hardware and software is actually manufactured, however, there are a number of threats that may still undermine the design:

Bugs or Vulnerabilities in the TCB—By definition, the TCB is fully trusted and assumed to be bug and vulnerability free. This is usually not explicitly stated, but should still be kept in mind while discussing security of any given architecture. Software (and hardware) design security verification [52] are themselves large research areas that can inform assumptions about the dangers of bugs and vulnerabilities in the TCB, and how to avoid them.

Hardware Trojans and Supply Chain Attacks—With the growing globalization of the supply chain, a single processor may include intellectual property (IP) blocks from multiple vendors from different countries, the whole system may be manufactured and processed in many different facilities in many countries before the final product is delivered to the customers. All the different parties who supply IP blocks or are part of the manufacturing and supply chain can potentially alter or modify the design to insert hardware trojans. When discussing the secure processor architecture design, it is usually implied that as long as the design is correct, the actual processor will be properly manufactured according to the design and not altered. Hardware trojan detection and prevention [231] as well as supply chain issues [178] are themselves large research topics that are well studied, and they can inform the assumptions about the threats of hardware trojans or possible supply chain attacks on the TCB.

Physical Probing and Invasive Attacks—Once a processor is manufactured and used in a real device, the device can be potentially easily probed through physical means. Some architectures may assume no physical attacks (e.g., processor is used in a system that is located in a secure facility) or they may assume limited physical attacks (e.g., the memory chip can be probed as it is separate from the main processor, but the processor itself cannot be probed). Typical physical attacks may involve reading out data from device via standard interfaces (e.g., remove DRAM chip, place in another computer, and read out contents of the DRAM memory chip [86]). They may also involve invasive attacks, such as decapsulating the package of the processor or memory to get access to the circuits on the chip. Such invasive attacks may use etching or drilling and use an optical microscope and a small metal probes to inspect the circuits. More sophisticated attacks can use focused ion beam (FIB) for probing of deep metal and polysilicon lines on the chip, or they can even be used to modify of the chip structure by adding interconnect wires or even creating new transistors [92]. Different secure processor architectures typically aim to protect against a subset of these attacks, e.g., assume DRAM memory can be removed and probed. There is a large body of ongoing research relating to physical attacks [213] that can be used to contextualize the assumptions about physical probing and attacks.

2.3   BASIC SECURITY CONCEPTS

Analyzing and designing a secure processor architecture involves deciding about what properties the system will provide for the protected software, within the limitations of the threat model. The basic properties are: confidentiality, integrity, and availability. Furthermore, authentication mechanisms are important to consider. As part of integrity checking and also of authentication, understanding freshness and nonces is a crucial aspect. Finally, it is important to distinguish security from reliability, and keep in mind that security assumes reliability is already in place.

2.3.1   CONFIDENTIALITY, INTEGRITY, AND AVAILABILITY

One of most basic objectives of any computer security system, whether hardware or software, is to protect code and data stored or executing on the system. There are three well-known security properties of the code or data with regard to which the code or data can be protected: confidentiality, integrity, and availability. These properties are defined by [128] as follows.

- "Confidentiality is the prevention of the disclosure of secret or sensitive information to unauthorized users or entities."

- "Integrity is the prevention of unauthorized modification of protected information without detection."

- "Availability is the provision of services and systems to legitimate users when requested or needed."

Confidentiality of code or data can be ensured if there is no action, or set of actions, that an untrusted entity can make to directly read, or otherwise deduce, contents of the confidential code or data. In terms of information flow, confidentiality can be modeled as existence of channels for information flow from the trusted entities to the untrusted entities. Such channels could be modeled as noisy channels or lossy channels if the untrusted entity only gets partial information about the confidential code or data. If the attacker is able to find a confidentiality breach, then the attacker can learn some or all information about sensitive code or data that the architecture aimed to protect, and the architecture in question needs to be redesigned and fixed.

It should be stressed that even partial information lean can be dangerous. Often attackers will have access to some external information or an ability to brute-force and make educated guesses once they have some partial information. The external information can be public information known to everybody, or some information attacker has obtained (e.g., through a different attack). This can help deduce information (e.g., combine known memory layout of a program with timing information for the cache). The ability to brute-force and check all possibilities, especially when guessing cryptographic keys, means that attackers do not actually need to get all the 128 bits for 128-bit AES, for example, they may get 96 or even 64 bits, and try to guess the rest in matter of days on a powerful computer. Finding an attack that requires no specific outside information, or no brute-forcing to work, is much more powerful and damaging than finding a very specific attack that only works in certain scenarios with lots of external information or requires brute-forcing some information.

Integrity of code and data can be ensured if there is no action, or set of actions, which allow untrusted entity to modify the protected code or data. Integrity attacks do not require an attacker to learn any information, but only the ability to modify something. Integrity attacks focus on code or data related to authentication or integrity checks, so as to allow the attacker to bypass these checks, and breach the system.

Integrity attacks can vary in the amount of modification the attacker attempts to do. On one end of the spectrum, an attacker may want to modify just one bit of information, e.g., enable/disable protections bit, that will allow him or her to later damage the system through a different attack. On the other end of the spectrum, he or she may try to modify whole memory contents to rewrite some code or data in the system.

Availability of the code and data can be ensured if there is no way for an attacker to deny service to the users of the system. Availability almost never can be achieved through use of one single secure processor design, e.g., the attacker can always smash the processor with a hammer to destroy it and thus deny it access to anybody. In less drastic approaches, attackers can slowly use up memory or other resources of the system, making it impossible for the protected code and data to execute in reasonable time. Availability can be achieved by using many secure processor working together, such as through redundancy.

2.3.2 AUTHENTICATION

Authentication relates to determining who a user or system is [128]. One approach to implementing authentication is for the parts of a system, or for a user and a system, to exchange information something that each knows, such as a password. This is usually referred to as proving "what you know," other approaches use proving "what you have" or "what you are" [128]. Authentication inherently requires integrity, as an attackers should neither be able to modify the authentication information nor make up their own.

2.3.3 FRESHNESS AND NONCES

When dealing with integrity or authentication, it is not only important that the information is correct but that it is "fresh." A good example of need for freshness are replay attacks where some previously correct and good data or information is re-sent at a future time when the information is no longer up to date. To ensure freshness, nonces are used. A nonce is a "number used once" and is a common way to ensure freshness in cryptographic protocols. Often, there is not a trusted, global clock that a components of a system can reference to find out if some event has happened before or after some time. An alternative is to use some indicator, the nonce, which can be referenced. As each value of the nonce is used once during a lifetime of the system, once a number has been used, it can be remembered. A practical way to implement nonces is to use a monotonic counter, thus only one, latest, value needs to be stored securely for reference by each component. An alternative could be to use random numbers as nonces (e.g., sender sends a random n and receiver replies with $n + 1$). When using random numbers as nonces, there is danger of two random number repeating themselves. Consequently, it may be best to use monotonic counters as nonces when it is possible to store the state (i.e., store the last nonce value."

2.3.4 SECURITY VS. RELIABILITY

From a security perspective, confidentiality, integrity, and availability assume a sophisticated attacker who attempts to maximize their chance of breaking the system. Security assumes that reliability, i.e., protection from random faults or errors, is already provided by the system, and focuses instead on the deliberate attacks by a smart adversary. Thus, reliability is about random errors, e.g., cosmic rays striking DRAM and causing a fault, while security is about deliberate attacks, e.g., attacker modifying exactly the memory location storing the secret key. When considering system availability, if the reliability is not maintained (e.g., the system shuts down), it still should not allow disclosure of any information to a potential attacker, nor allow information to be modified.

2.4   SYMMETRIC-KEY CRYPTOGRAPHY

To ensure data confidentiality, symmetric- or private-key cryptography is needed. In symmetric-key cryptography, data encryption and decryption uses the same secret key. When protecting data, there is the plaintext p which is encrypted into a resulting ciphertext c by the encryption function which also uses some private key k. Thus, the encryption process is: $c = Enc(k, p)$. To get back the plaintext data, decryption is needed: $p = Dec(k, c)$. Note, both encryption and decryption use the same key in symmetric-key cryptography. It is required that the encrypted ciphertext looks almost random to someone who does not posses the key k. Given ciphertext, an attacker should not be able to learn neither the key nor the plaintext data. Symmetric-key algorithms can be broken down into block ciphers and stream ciphers.

2.4.1   SYMMETRIC-KEY ALGORITHMS: BLOCK CIPHERS

Block ciphers work on blocks of data, e.g., AES uses a 16-byte block. Plaintext to be encrypted has to be a multiple of the block size. Data smaller than the block needs to be padded to the block size, while data bigger than the block size is encrypted in block-sized chunks. For data bigger than one block size, there are different modes of operation of the block cipher that determine how the blocks are encrypted [128]. Electronic Code Book (ECB) mode simply encrypts one block at a time. The danger of ECB is that encryption of same data blocks will result in same ciphertext blocks when using same keys, which can reveal patterns about the input plaintext data. Other modes include Cipher Block Chaining (CBC) or Counter Mode (CTR). Here, the same input data blocks will results in different ciphertext blocks, even when using same key.

Encryption only provides confidentiality, but integrity is often also desired. For integrity protection, there are dedicated modes of operation [128], e.g., Hash-based Message Authentication Code (HMAC), Cipher-based Message Authentication Code (CMAC), or Galois Message Authentication Code (GMAC). In these modes, the last ciphertext block is effectively a secure hash (fingerprint) of the whole data, and its value depends on the encryption key.

To avoid having to separately do encryption and hashing, there are authenticated encryption modes which combine the two operations into one. One of the most recent recommended modes is the Galois/Counter Mode (GCM) [148]. Given a key, it can encrypt data and generate a keyed-hash value of the data.

Decryption works similar to encryption, where one has to work with block-sized chunks of ciphertext. Some modes of operation, e.g., CTR, allow for easy decryption of random blocks inside the ciphertext. With CTR, each block is *xor*ed with an encryption of a counter, knowing which counter corresponds to which block, a random block can be decrypted by encrypting the counter value and *xor*ing it with the ciphertext block. Meanwhile, others may require to decrypt multiple blocks together or a whole ciphertext (due to the chaining, in a serial fashion, among the individual blocks).

2.4.2   SYMMETRIC-KEY ALGORITHMS: STREAM CIPHERS

A different approach to symmetric-key encryption is taken by stream ciphers. Here, the plaintext is encrypted bit by bit by combining it (using *xor* operation) with a pseudorandom keystream. The keystream is typically generated from a random seed using shift registers, and the seed is the cryptographic key needed for decryption. Some modes of operation of block ciphers behave as stream ciphers, but still the main distinction is that they operate on blocks of data, while stream ciphers operate on bits.

Stream ciphers can be faster in hardware than block ciphers. One disadvantage is that is hard to provide ability to do random data access inside the encrypted ciphertext as one has to re-generate the pseudorandom keystream up to the point where the to-be-access data is located in the ciphertext stream.

2.4.3   STANDARD SYMMETRIC: KEY ALGORITHMS

While a number of symmetric-key algorithms exist, designers should use the AES [173] which is well studied and considered secure. There are also some "lightweight" block ciphers, such as PRESENT [27]. Older algorithms such as RC4, DES, or 3DES should no longer be used as their key sizes are too small or they are considered insecure due to cryptographic attacks. Moreover, custom-designed algorithms should be avoided. There are numerous examples of security-by-obscurity where custom, publicly untested algorithms have been deployed in products such as metro cards, only to be found to have serious flaws [44]. Among stream ciphers, there are Salsa29 and its variant ChaCha [22].

2.5   PUBLIC-KEY CRYPTOGRAPHY

In public-key cryptography, also called asymmetric-key cryptography, data encryption and decryption uses different keys. For confidentiality protection, the input is a plaintext p which is encrypted into a resulting ciphertext c by the encryption function which uses the public key pk. Thus the encryption process is: $c = Enc(pk, p)$. To get back the plaintext data, decryption is needed: $p = Dec(sk, c)$. Here, the decryption uses the secret key sk. Given a pk it should be infeasible to find out what is the secret key sk, which depends on hardness of certain mathematical problems, such as factoring of large numbers, e.g., RSA [175]. The advantage of public-key cryptography is that pk can be given to anybody, and they can encrypt the data or code using this pk. Meanwhile, only the user, program, or hardware module in possession of the sk can decrypt the data.

For integrity, public-key cryptography can be used in "reverse" direction when used in digital signatures or message authentication codes. The sk can be used to create a digital signature, and anybody with access to the *pk* can verify the signature—but cannot make a new valid signature as they do not have *sk* nor can they get the *sk* from knowing *pk*.

2.5.1   KEY ENCAPSULATION MECHANISMS

Key Encapsulation Mechanisms (KEMs) are methods for securely transmitting a symmetric encryption key using public-key encryption. Public-key encryption is typically significantly slower and more costly in terms of computations. As result, in most applications which require encryption of a lot of data in a public-key setting, the encryption key is secured and transferred using the (relatively slow) public-key encryption, while the actual data is later encrypted (relatively fast) using that symmetric key. Upon receiving a message, first the public-key algorithm is used to decrypt the key, then the symmetric-key algorithm is used to decrypt the actual data.

2.5.2   STANDARD PUBLIC-KEY ALGORITHMS

The most well-known public-key algorithm is the RSA algorithm [175], which derives its security from hardness of problem of factoring large numbers. More recently, Elliptic Curve Cryptography (ECC) [87] has gained popularity as well, which is based on the algebraic structure of elliptic curves over finite fields.

2.5.3   POST-QUANTUM CRYPTOGRAPHY

Most recently, there is active interest in the so-called Post-Quantum Cryptographic (PQC) algorithms, as the promise of practical quantum computers nears [37]. In the 1990s, Shor proposed algorithms that can solve both the integer-factorization problem and the discrete-logarithm problem in polynomial time on a quantum computer [197, 198]. Cryptosystems based on the hardness assumptions of the integer-factorization problem and the discrete-logarithm problem can be broken using Shor's algorithm and a quantum computer. For example, public-key algorithms such as RSA may not be secure for long-term use and other algorithms need to be standardized and used in their place.

As of 2018, there are five categories of mathematical problems that are under investigation as candidates for PQC: code-based systems, lattice-based systems, hash-based systems, systems based on multivariate polynomial equations, and systems based on supersingular isogenies of elliptic curves [23, 179].

In addition, Grover's algorithm [78] gives a square-root speedup on brute-force attacks that check every possible key. This does not break symmetric-key algorithms, such as AES, but does necessitate the use of larger keys. For example, a 256-bit pre-quantum security level corresponds to 128-bit post-quantum security level.

2.6   RANDOM NUMBER GENERATION

Most security and cryptographic algorithms and protocols depend on good sources of random numbers, especially for key generation. There are pseudo-random number generators (PRNGs), which use an algorithm to expand a seed into a long string of random-looking numbers. The numbers are not truly random, as given knowledge of the seed and the algorithm; anybody can re-generate the same sequence of random-looking numbers. There are also true random number generators (TRNGs), which generate truly random numbers. For example, physical phenomenon like electrical noise or temperature variations can be used as sources of randomness. True random numbers are hard to obtain at high rate, thus many times TRNGs generate a small true random number, the seed, and a PRNG is used to expand that seed into a long string of random numbers. As long as the seed is truly random, and never accessible to potential attackers, then the resulting PRNG output can be used in secure manner. Computer architects often assume existence of sufficient randomness, and hardware and circuit designers are ones focusing on how to create such circuits. Designers should however realize that TRNGs can be manipulated, such as by inserting backdoors into processor's random number generator [166].

2.7   SECURE HASHING

To ensure data integrity, secure hashes are needed. A secure hash algorithm is a cryptographic hash function. A secure hash maps input data, m, of variable size to a fixed size output, h—the output is simply called the hash of the input data. Secure hash is a one-way function, and it should be infeasible to mathematically invert it and deduce the input data given the hash value. Furthermore, there are three properties that strong cryptographic hashes should have. They are as follows.

- Pre-image resistance—given a hash value h it should be infeasible to find any message m such that $h = hash(m)$.

- Second pre-image resistance—given a specific input m_1 it should be infeasible to find different input m_2 such that $hash(m_1) = hash(m_2)$.

- Collision resistance—it should be difficult to find two random messages m_3 and m_4, where $m_3 \neq m_4$, such that $hash(m_3) = hash(m_4)$; due to the birthday paradox [128] it is possible to find two such random messages that hash to the same value much more readily than one may expected.

Hash functions are used to compute the hash value, sometimes called digest or fingerprint, of some input data. Given the hash size is fixed, and much smaller than the data size in most cases, it is easier to store and protect the hash value, rather than the original data. Given same input m, the hash will always be the same h, and anybody with access to m can compute h.

Common application of hashes is in authentication and integrity checking. For example, a hash can be computed for a large file, then the file can be sent to an untrusted storage while the hash is kept in a safe location. Later, when the large file is read again, its hash can be recomputed and checked against the stored value to make sure there was no modification to the file (note this does not protect against replay attacks, or if someone is able to change the hash value stored in the secure location). When checking integrity or authentication, freshness needs to be ensured; often a nonce is included as part of the hash (i.e., hash data concatenated with the nonce).

2.7.1   USE OF HASHES IN MESSAGE AUTHENTICATION CODES

When using Message Authentication Codes (MACs) only the entity with the correct cryptographic key can generate or check the hash value. Unlike secure hashes where anybody can generate the hash value, MACs have added requirement that only select entities who have the correct cryptographic key can do so. The advantage of MACs over plain secure hashes is that the hash value cannot be generated by anybody, just the entity that has the key. MACs can be sent over untrusted communication channels along with the data. The attacker may try to change the data, but he or she cannot generate a new MAC that matches the data as long as he or she does not have the cryptographic key. To check the integrity, data can be hashed again by the receiver, and then hash compared with the decrypted hash from the MAC. MAC can be realized by using keyed-hash message authentication codes, block ciphers, or based on universal hashing. Both sender and receiver need to share the same key when using MACs.

2.7.2   USE OF HASHES IN DIGITAL SIGNATURES

Digital signatures are similar to MACs, in that only the user or program with the correct cryptographic key can generate or check the hash value. Unlike MACs, digital signature use a private signing key to generate the signatures, and a public verification key to check the signatures. Akin to public-key cryptography, the parties generating and verifying the digital signatures have different keys. Digital signatures can leverage public-key cryptographic algorithms, e.g., on the sender's end securely hash a message then encrypt the hash with the private key which allows the verifier to re-generate the hash, on the receiver's end decrypt the received value using the public key, and check if the two match. The public key infrastructure can be leveraged to distribute certificates so that sender and receiver need not to have direct contact in order to be able to authenticate the messages that were digitally signed.

2.7.3   USE OF HASHES IN HASH TREES

A hash tree, also called a Merkle tree [150] and shown in Figure 2.2, is usually a binary tree data structure. In the hash tree, a parent node contains hash value of its children nodes. At the very bottom of the tree are the leaf nodes. Thus, as one proceeds up the tree, integrity of the lower tree nodes can be checked by computing their hash value and comparing with the hash stored in the parent nodes; finally the root node contains the hash value which is dependent on the values of all the intermediate nodes, and thus the leaf nodes.

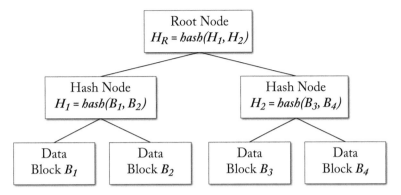

Figure 2.2: Example of a hash tree with four data nodes, showing internal hash nodes, and the root hash. The value of the root hash depends on all the data nodes' values.

The hash tree was invented to speed up the verification of integrity of large sets of data. Assuming leaf nodes correspond to parts of a file, or parts of the computer memory, if there is a change in part of the file, or part of the memory, one only needs to check the hash value of the parent nodes of the leaf nodes that were changed—and not compute the hash over the whole file or memory.

2.7.4 APPLICATION OF HASHES IN KEY DERIVATION FUNCTION

Often, a system uses many symmetric keys, or needs to generate a symmetric key in a specific format. Key Derivation Function (KDF), is used to derive one or more secret keys from a master secret key. Keyed cryptographic hash functions are one example of pseudo-random functions that can be used for key derivation [32].

2.7.5 STANDARD SECURE HASH ALGORITHMS

There exist numerous secure hash algorithms. Most common, and most recommended, ones today are the SHA-2 [202] and SHA-3 [57] algorithms. Use of older algorithms such as MD4, MD5, or SHA-1 show be avoided as they are no longer secure as a commodity computer can be used to perform a brute-force search to find a message that gives the desired hash value. A hash digest size of 256 bits is likely a good choice today as it makes it impossible launch a brute-force attack, even with a quantum computer (due to Grover's algorithm, pre-quantum security level of 256 is equivalent to post-quantum 128-bit security level).

2.8 PUBLIC KEY INFRASTRUCTURE

Public key infrastructure (PKI) is a set of policies and protocols for managing, storing, and distributing certificates used in public-key encryption [14]. In PKI, there is a trusted third party which can distribute digital certificates that vouch for correctness of public keys pk of different entities, and allows for verification and decryption of public-key encrypted data without having to directly talk to each sender to get their key.

The trusted third party is the certificate authority. The authority is responsible for verification of the certificates it receivers. It then distributes the certificates to other users, signed with its own private keys. Certificates for the certificate authorities are usually pre-distributed (e.g., browsers come with built-in list of certificates for certificate authorities). If there is a problem with a certificate authority, then the PKI infrastructure will break. For example, a user can by mistake trust a certificate authority they should not. Or, a malicious certificate authority could equivocate and give different information to different users, enabling man-in-the-middle type attacks, for example.

2.8.1 DIGITAL CERTIFICATES

Digital certificates are a central part of the PKI. In a simplified form, a digital certificate contains some identifying information about a system or a user and their public key. This information is encrypted with a private key of the certificate authority. Users are given known-good public keys of the certificate authorities, which they can use to check authenticity of the certificates. Once authenticity of certificate is verified (i.e., the digital signature of the certificate was indeed made by the trusted authority) then the users can be sure that the public key in the certificate belongs to the system or user that the certificate is for. With this information, users can verify messages sent by the entity that has the private key corresponding to the public key in the certificate.

2.8.2 DIFFIE–HELLMAN KEY EXCHANGE

Diffie–Hellman (DH) is an algorithm used to establish a shared secret between two entities. Its main application is as a way of exchanging cryptography keys for use in symmetric-key algorithms. Symmetric-key encryption and decryption is much more efficient than public-key encryption. Thus, often it is desired to use public-key protected means of communication to generate a shared, symmetric secret key for actual transfer of data between two entities. Diffie-Hellman requires use of authenticated channel, i.e., the two communicating parties need to authenticate each other, otherwise there is a man-in-the-middle attack.

2.8.3 APPLICATION OF PKI IN SECURE PROCESSOR ARCHITECTURES

As will be seen in later chapters, a PKI is not directly used by secure processor hardware. However, for users of such processors, a PKI is needed to distribute certificates vouching for the cryptographic secrets embedded in these secure processors (such as a private key pk of the processor). Each secure processor typically needs at least one private key (either burned-in in some registers, stored on non-volatile memory, or derived from a PUF). To ensure users that they are indeed communicating with the expected processor, the manufacturer typically needs a PKI to distribute certificates vouching for the public keys assigned to each processor.

Having a publicly known signing key may be a privacy issue, allowing adversaries to de-anonymize communication if the see the same key being used over and over again. This issue has been addressed by work such as on Direct Anonymous Attestation (DAA), which is a cryptographic primitive used to enable remote authentication of a trusted computer while preserving privacy of the platform's user [29].

2.9   PHYSICALLY UNCLONABLE FUNCTIONS

Each secure processor should be uniquely identified and have a unique set of cryptographic keys. This can be achieve by "burning in" the unique information at the factory. However, such an approach requires extra cost (since each chip has to be written with the unique information), and a potentially malicious manufacturer may keep information about all the secret keys they have burned into their products. As an alternative, researchers have recently proposed PUFs [143], and they could be used to generate unique information per-chip.

PUFs leverage the unique behavior of a device, due to manufacturing variations, as a hardware-based fingerprint. A PUF instance is extremely difficult to replicate, even by the manufacturer. Many uses of PUFs have been presented in literature: authentication and identification [119, 206, 224], hardware-software binding [81, 82, 123, 186, 187], remote attestation [125, 189], and secret key storage [225, 226]. PUFs can also be leveraged for random number generation [144] as well as for recently proposed virtual proofs of reality [182]. PUF implementation is mostly domain of hardware and circuit designers, but architects can leverage them in their security architectures. At the architecture level, PUFs are most often abstracted away as modules that give a unique, unclonable fingerprint of the hardware.

CHAPTER 3

Secure Processor Architectures

This chapter introduces the main features of secure processor architectures, as an extension of general-processor architectures. It first gives examples of real-world attacks on existing processor architectures, which further motivates research and design of secure architectures. It next presents features of general-processor architectures. It then introduces secure processor architectures, their features, especially concerning different execution levels and privileges, and gives examples of existing academic and commercial architectures. The chapter also lists secure processor architecture assumptions and limitations of what the architectures can achieve.

3.1 REAL-WORLD ATTACKS

Before diving into description of secure processor architectures, this section gives some examples of-real world attacks that modern processors face. These types of attacks further motivate the need for secure processor architectures, and also give a warning of how wrong assumptions about the hardware behavior (e.g., the decay of DRAM after power is turned of), or unintended consequences of performance improving features (e.g., speculative execution), can result in vulnerabilities and attacks.

As secure processor architectures are built upon normal processors, the same bugs and vulnerabilities can affect them as well, and should be guarded against. These vulnerabilities are in addition to any intentional hardware trojans or modifications to the system. The processor vulnerabilities can be in the ISA, microarchitecture, circuits, or devices, and they can be used to break the system or bypass protections offered by the system. The attacks range from ones that simply crash the system, to attacks allowing one to steal data from different processes, kernel, or form other VMs. Some of the attacks can be deployed remotely without physical access to the system, only requiring the attacker to run some code on the target machine, while other attacks require physical presence, e.g., to probe the circuits or remove DRAM chips.

The canonical example of a hardware processor bug (although not security related) is the FDIV bug in Intel processors, discovered in 1994 [170]. The Pentium FDIV bug in the Intel P5 Pentium Floating Point Unit (FPU) caused incorrect decimal results to be returned for floating point division. While the impact of the bug on average users maybe unclear, it resulted in hardware recalls and millions of dollars in costs to Intel [170].

Of the security-related hardware bugs or vulnerabilities, the recent and most well-known ones are the Coldboot attack [86] and the Rowhammer attack [117], both affecting DRAM. There are also the major processor vulnerabilities from 2018 published as this book was being written: Spectre [120] and Meltdown [138]. These vulnerabilities and resulting attacks exemplify the range of threats that processors face, from ones requiring physical access to cool down the memories to later steal data, to ones that can be executed remotely on cloud computing servers. They also show that any component in the computer system can be vulnerable, e.g., a problem with the DRAM may be just as damaging as problem with the main processor chip itself.

3.1.1 COLDBOOT

A Coldboot [86] attack can be used to steal information from DRAM when the system is powered off. Coldboot exploits physical phenomenon that data stored in DRAM does not disappear as soon as the power is turned off. Rather, with the DRAM refresh disabled, or the power all-together turned off, the charges on capacitors in the DRAM cells (which are used to store the data) slowly decay. Thus, the basic assumption that DRAM is a volatile memory that looses contents instantly when powered off is not true.

In the Coldboot attack, researchers have shown that data, such as encryption keys, can be extracted from DRAM chips after computer is powered off. To extend the amount of time available before charges in the capacitors decay, DRAM chips can be easily cooled down with a can of compressed air spray (kind of used to clean computer keyboards or other electronics from dust). Cooling DRAM further slows down the decay, allowing one to remove a DRAM module from the computer, transfer it to another computer and dump the data, for example. Alternatively, a computer can be quickly shut down while DRAM is cooled, and rebooted into a malicious OS that reads off the DRAM data before it could have decayed.

A variety of solutions can be used to protect against Coldboot attack, but all require extra software or hardware to effectively erase data, rather than wait and assume the data will be lost due to the DRAM cell decay. In software, secret keys need to be explicitly erased. In hardware, battery-backed DRAM could use stored energy from the batteries to explicitly zero out the memory contents when external power is lost or refresh is disabled.

3.1.2 ROWHAMMER

A Rowhammer [117] attack can be used to alter bits in memory locations not accessible to the attacker process or application. Rowhammer is a different vulnerability of the DRAM, but one which is also related to how data is stored as charges on capacitors in DRAM. Most computer systems rely on isolation to separate programs or VMs from one another. The isolation is enforced through page tables or other mechanisms for checking which physical memory a process can access. However, as shown in Rowhammer, accessing DRAM cells in a specific pattern can cause data to be altered in other DRAM cells—there is no explicit violation of the isolation mechanisms, but rather the physical devices' properties cause data to change in memory locations what were not actually accessed by the attacker.

To realize the attack, first, attacker process' data and the victim's data need to be in adjacent DRAM rows. Next, the attacker can repeatedly access its own data in the DRAM rows adjacent to the victim's data. After a large number of iterations, some of the bits in the victim's data will change their value. The attack is built on the principle that the charges in certain DRAM cells will flip if there is repeated electrical activity near by, i.e., memory access, in the adjacent cells. Which cells flip their value depends on the manufacturing variations of the DRAMs and is different from one device to another. Thus, the attack requires the data to be in very specific locations, and not all DRAM rows in a DRAM module may be susceptible to this attack.

Protections against this type of attack can include hardware modifications to make DRAM cells less prone to flipping bits under repeated stress of accesses to adjacent cells. In software, the memory of victim and attacker processes can be allocated such that it is not in adjacent DRAM rows.

3.1.3 MELTDOWN

Meltdown [138] vulnerability can be used to break isolation between user applications and the operating system. Meltdown exploits side effects of the out-of-order and speculative execution features on today's processors to enable user applications to read arbitrary memory locations of the operating system kernel that have been mapped into the address space of the user process. Today's operating systems map the kernel into the address space of every process to allow for, e.g., fast interrupt handling that does not require changing address spaces. The isolation between the user application and the kernel memory locations is based on a privilege level bit indicating the current execution privilege level (typically user or kernel). On a memory access, the privilege level is checked. The out-of-order and speculative execution features of Intel processors, but possibly others as well, allow for speculative execution instructions to improve performance; and they should nullify any changes in the processor state if the speculation was incorrect. However, speculative execution of data loads also influences the processor cache. Although the processor my properly clean up its state after any speculative execution, if the speculatively loaded data remains in the processor cache, it can leak information, as was demonstrated with Meltdown [138] (and related Spectre attack discussed shortly).

In a simplified example of Meltdown, there may be an instruction that causes a trap in a user program, followed by an access to a memory location in the kernel, labeled *data* in below example, and further access, labeled *probe_array* in below example, that uses the data from that kernel memory location as an address to access another memory location. A sample code form [138] is:

```
raise_exception();
access(probe_array[data * 4096]);
```

If the out-of-order and speculative execution logic speculatively executes the memory access (e.g., before computing that the instructions should not happen due to the trap), it will read the data from kernel memory location and use it as an address for the probe array load. The accessed kernel data is never visible to the user application. However, the memory that was accessed based on the address derived form the data in the kernel memory is left present in the cache.

Subsequently, by doing a cache side-channel attack [138], the attacker application can probe which memory locations are in the cache, and from there can directly derive the value of the kernel's data. Thus, out-of-order and speculative execution, combined with a cache side-channel attack, allow user applications to bypass protection checks and read any data that is mapped into the user application's address space.

A hardware solution to Meltdown is to do privilege checks on the speculatively executed data early in the speculation process (processors such as from AMD were not found to be vulnerable as they do the checks before the memory access is speculatively executed). Such hardware changes require micro code updates or replacement of the processor which can be costly. A software solution is to not map so much kernel data into the address space of user applications, however, this will have performance impact, e.g., on interrupts.

3.1.4   SPECTRE

Spectre [120] vulnerability can be used to break isolation between different applications. Spectre exploits speculative execution of instructions following branch instructions. Compared to Spectre, Meltdown does not use branch prediction for achieving speculative execution; it relies on instructions that will cause a trap. Also, Meltdown leverages delayed privilege checks to allow applications to access kernel memory locations—meanwhile Spectre allows forcing a victim application to leak its secrets to a different, attacker, application. Both vulnerabilities leverage cache side-channels attacks to actually find out what is the secret data (of the kernel in case of Meltdown, or of the victim application in case of Spectre). According to the researchers who found Spectre vulnerability, speculative execution capabilities found in processors from Intel, AMD, and ARM all currently have Spectre vulnerability [120]. Consequently, Spectre affects most processors in use today, but in practice may be difficult to exploit as it requires a mix of techniques to achieve a practical attack.

In a simplified example, an attacker needs to train a branch predictor to mis-predict on a certain branch instruction address. As a branch predictor is shared by all processes running on the same CPU, an attacker who knows the code of the victim can create an application that has branches at same addresses, but the branch outcome is different. Furthermore, the attacker should flush the cache. For the victim, as an example, it can have an access to an array, result of which is used to access a second array; with an *if* statement to guard the address range for the first array access—to prevent accesses beyond its bounds. A sample code form [120] is:

```
if (x < array1_size)
    y = array2[array1[x] * 256];
```

Since the attacker has trained the branch predictor to mis-predict on this *if* branch check, it can cause the processor hardware to assume the branch is not taken (*if* statement is true) and execute the out-of-bounds access with a very large x to the first array, which will be used by the second array access. The processor will eventually compute that there was a mis-prediction and discard any data, y. However, the processor cache will have already been occupied with the data from array accesses. As the array element accessed depends directly on value of data in *array*1[*x*], in a final step, the attacker can perform a cache side-channel attack and measure timing of the accesses to figure out which memory location was brought into the cache during the speculative execution [120] and thus find the value of the data at *array*1[*x*]. In addition to abusing the branch predictor, other variants of the attack are possible such as by using indirect branches [120].

A hardware solution for this type of attack would be to disable the speculative execution and branch prediction. However, the performance impact would be significant. Another hardware option would be not to share the branch predictor (e.g., have multiple separate predictors in hardware), but such solution may not scale. A software solution may be that applications could also be isolated by having one application only running alone on one processor. Since processors do not share branch predictors, this would prevent an attacker from influencing the branch predictor state—but again performance impact would be significant. Another, partial, software solution is to for loads inside branches to act as memory fences (or insert explicit memory serialization instructions). This can prevent memory loads (which modify cache state used in the side channel part of the attack) from executing until prior instructions have finished.

3.1.5 OTHER BUGS AND VULNERABILITIES

Although some of the bugs or vulnerabilities attract much attention in the popular media, there are many more that affect, or have affected, computer systems. Normal processors suffer from variety of bugs, which are published regularly by the processor manufacturers in their errata documents, or which are listed on numerous web pages, e.g., [45]. An analysis of over 300 bugs from these errata documents has found almost 10% were security-critical [94]. Attacks leveraging processor bugs, or simply abusing some processor functionality, can be used against features such as the System Management Mode (SMM), to escalate privileges of the attacker [237], or Message Signaled Interrupts (MSI) mechanisms to break VM isolation [238]. The vulnerabilities are by no means limited only to processors or memories, as shown above. For example, researchers have demonstrated that vulnerabilities in GPUs can be used to break isolation and steal data from different programs sharing the same GPU [130].

Attacks do not have to also focus just on the compute related components. Thermal sensors have been abuse to leak information in multicore processors [19]. Features such as dynamic voltage and frequency scaling (DVFS) have also been abused, and researches have shown that they can be manipulated to change timing of operations and introduce faults, allowing them to leak secrets such as encryption keys from protected environments [210].

Many of such attacks stem from competing design goals. The main goals of processor architects are to improve performance, reduce area, or reduce energy. Each type of optimizations, however, can bring about potential vulnerabilities. Performance enhancing features are a prime example of sources of potential attacks, for example, processor caches allow for creating abstraction of large, fast memory, but they also allow cache side-channel attacks due to timing differences in memory accesses that they create. Reduction of area can also lead to potential attacks, e.g., using more densely packed transistors and electronics can reduce cost of a chip, but can lead to attacks such as Rowhammer in DRAMs, where close proximity of memory cells allows them to interact electrically in ways that break higher-level assumptions about how the memory operates. Power features such as the dynamic voltage and frequency scaling can cut processor power, but also can be abused to change timing of operations and cause faults that result in attacks.

3.2   GENERAL-PURPOSE PROCESSOR ARCHITECTURES

Secure processors are built on top of general-purpose processor architectures, and expand them with new security features. A diagram of the typical software levels in a modern processor is shown in Figure 3.1. On the left-hand side are shown the typical protection levels, or "rings." At any point in time, code running on the system is executing at one of the privilege levels, or rings. The privilege level determines what the code or instructions can and cannot do.

Traditionally, modern computer systems use ring-based protections, first proposed in hardware around 1970s [188], to separate privileged and unprivileged software. During code execution, hardware keeps track of the current privilege level (ring) in which the code is running. The different rings considered in commodity processors are: user (ring 3), various semi-privileged code (ring 2 and 1), and the operating system kernel (ring 0). To provide further functionality, over time, new features have been added, resulting in addition of new privilege levels, such as the hypervisor (ring -1).

When executing instructions on a processor, certain instructions are restricted to be only available in privileged mode (e.g., ring 0) while others can be executed in any privilege level. Also, memory access checks are performed to ensure only privileged code can access some memory locations, for example. A privilege change from a lower to higher privilege can happen through special instructions (such as a system call initiated by the application or a VM exit initiated by

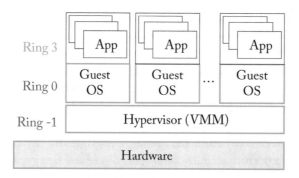

Figure 3.1: A diagram of the typical software levels in a modern processor. The green outline shows the components most often considered trusted in a modern processor.

the guest VM) or by a hardware event (such as a fault, an interrupt, or a signal on a physical pin on the processor chip).

Entrance to more privileged modes has to be guarded so that less privileged code cannot elevate its privileges when not authorized. It is the duty of the software and hardware controlling the more-privileged execution to validate the inputs before it acts on them.

Any fault that occurs while code is in a particular privilege mode can affect all other code in that or any less privileged (i.e., same or higher ring number) code. Such faults, however, should not have impact on any more privileged code (i.e., lower ring number). For example, if a guest OS crashes (ring 0), then the hypervisor (ring -1) should still keep operating correctly.

As Figure 3.1 highlights with green outline, the operating system and the hypervisor are today trusted. Compromised or malicious operating system can attack all the applications in the system. Likewise, compromised or malicious hypervisor can attack all the operating systems in the system. Secure processor architectures, as discussed later, address some of these issues by adding new privilege levels for trusted management software, prevent some lower levels (more privileged today) from having access to higher levels, or add horizontal privilege separation within levels. This aims to help reduce the software TCB, which otherwise today contains all the software from the operating system down.

3.2.1   TYPICAL SOFTWARE LEVELS (RINGS 3 TO -1)

User (ring 3) to operating system kernel (ring 0) protections are well covered by standard computer architecture textbooks, e.g., [164]. Starting in the early 2000s, commodity processor vendors have started to add hardware virtualization assistance to their products, with Intel VT-x [227] and AMD-V [5] extensions. Like many ideas, however, virtualization support has been explored prior to that by mainframe computers such as from IBM or Honeywell [79] LPAR.

In today's commodity processors, hypervisor mode (ring -1) is typically implemented as an extra privileged mode for the main processor cores. Hardware changes only affect the processor. In addition to the mode itself, extra additions such as hardware nested page tables [24] have been introduced (a concept of software-based shadow page tables [5] has been used before hardware supported nested page tables were introduced). With addition of these components, it is natural that hardware inside the processor, such as table-lookaside buffers (TLBs), is expanded to work with the new memory management additions. For example, Intel processor include hardware page-table walker which can handle the extra levels of page tables [7]. Meanwhile, IBM's mainframes have also supported nested virtualization before [160].

3.2.2   TYPICAL HARDWARE COMPONENTS

A typical computer system today contains the main processor, and other components such as memories or input and output (I/O) devices. All the components are connected through a memory bus. As Figure 3.2 shows with green highlighted outlines, all of the hardware components of a processor are trusted today. Information can be extracted from memory or memory contents can be modified. Snooping on the system bus is possible and can be used to extract information communicated on the bus. Compromised or malicious devices can attack other components of the system. Secure processor architectures, as discussed later, add new features to the processor, or the other components, so that some of the other components can be untrusted (and thus not part of the TCB).

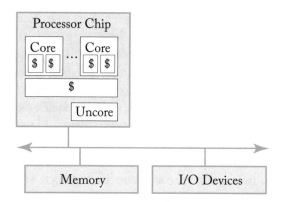

Figure 3.2: A diagram of the typical hardware components in a modern processor. The green outline shows the components most often considered trusted in a modern processors.

3.3   SECURE PROCESSOR ARCHITECTURES

Secure Processor Architectures add new hardware and software features to provide Trusted Execution Environments (TEEs) wherein software executes protected from some of the software and hardware threats (according to an architecture's threat model). Secure processor architectures enhance general-purpose processor with new protection features. They provide new or alternate privilege levels and utilize software or hardware changes to facilitate protection of software (software modules, applications, or even VMs).

3.3.1   EXTENDING VERTICAL PRIVILEGE LEVELS

One of the first extensions that can be made to general-purpose processor architectures is addition of new privilege levels, shown in Figure 3.3. The new privilege levels include system management mode (ring -2) or platform security engine (ring -3).

System Management Mode (Ring -2) is a privilege level originally introduced in Intel processors [103], which is more privileged than the hypervisor mode. The SMM code is typically part of the firmware, and the SMM mode can be entered only through a special System Management Interrupt (SMI) by asserting a pin on the processor chip's package or I/O access to a specific port.

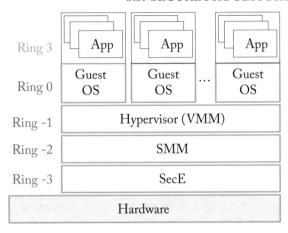

Figure 3.3: A diagram showing new privilege levels that have been added over time. New levels are leveraged to protect more trusted code, as it manages the system.

The goal of SMM to provide some management functionalities, even if the operating system or hypervisor is compromised. The SMM code is typically very small and provided by the computer manufacturer without means for users to analyze, check, or update the code. These restrictions create a security through obscurity situation, and have led to exploits, e.g., SMM rootkits [59].

Platform Security Engine (Ring -3) is an even more privileged level. Platform Security Engine is a name given in this book to solutions similar to Intel's Management Engine (ME) [181] or AMD's Platform Secure Processor [10]. Platform management engine is typically a separate, small processor fully independent of the main processor of the computer. The management engine can access resources of the computer even if the operating system or hypervisor is compromised or has crashed. If management engine is a separate chip, it can have separate power connection from the main processor, memory or other components allowing it to stay on while the whole computer is offline—and, for example, it can be used to power on computers remotely (management engine has interface to network as well so it can receive remote commands). In addition, it often has reserved memory regions that it can use for code or data, which cannot be accessed by other components, leading to the designation of ring -3 that is not controllable by any other code in the system. In case of Intel's ME, it was usually embedded into the motherboard's north bridge [183], although with newer designs it may be in another part of the system, or even within same package as the main processor. With introduction of AMD's SEV and memory encryption technologies, AMD's chips include a Platform Security Processor [10], which has many similarities to the Management Engine.

The goal of the highest privileged level is to be able to control system execution and emulate some hardware features using a very small, embedded processor. However, similar to SMM, the even more privileged security engine usually, in commercial products, contains proprietary code that is usually a trade secret, with infrequent updates. There are today no means for users to analyze, check, or update the code. This secrecy introduces type of security through obscurity, and there are recent revelations of bugs that may lead to attacks, for example, on the Management Engine in Intel [60].

3.3.2   HORIZONTAL PRIVILEGE LEVEL SEPARATION

New privileged execution modes can also be introduced to separate privileges horizontally. These new privileges can be made orthogonal to existing protection levels. For example, using ARM's terminology, a "normal" and "secure" execution modes can be created, where within each mode there are still the usual application, operating system, etc., levels. However, the software executing in the secure mode (regardless of the level) is more privileged than any software executing in the normal mode.

3.3.3   BREAKING LINEAR HIERARCHY OF PROTECTION LEVELS

Furthermore, architectures can be designed to break the linear relation ship (where the lower level is always more privileged than a higher level). Figure 3.4 shows examples of how different architectures have in the past been designed to omit some of the levels from the software TCB. Reducing the number of trusted levels makes the TCB smaller, which is always one of the design goals for secure processor architectures. In Figure 3.4, normal processor is shown with all the levels trusted (in green outline). Some architectures, such as Bastion [35], have been designed to assume untrusted operating system; hypervisor and all the levels below work together to protect the Trusted Software Modules (TSMs). Intel's SGX [13] is an example of architecture that removes most of the levels from the software TCB, and where the security engine is only one needed, and trusted, to protect an Enclave (equivalent of a TSM). AMD's approach with SEV [50] is to use the security engine to protect whole VMs, and all the code inside them.

3.3.4   CAPABILITY-BASED PROTECTIONS

There are also alternative designs which do not necessarily use the linearly ordered set of privilege levels at all. Capability-based architectures [239] associate "capabilities" with different resources (such as memory locations or hardware modules). To access or make use of a resource, the requester needs to have a token or a capability that allows it the access. In such scenarios, the threat models will look different as the trusted and untrusted components (and potential attackers) are not discussed in terms privilege levels, but sets of capabilities that each entity possesses.

3.3.5   ARCHITECTURES FOR DIFFERENT SOFTWARE THREATS

As discussed before, there are numerous software and hardware attacks that a system can aim to protect against. In Figure 2.1, the attack surface was shown and four types of attacks were

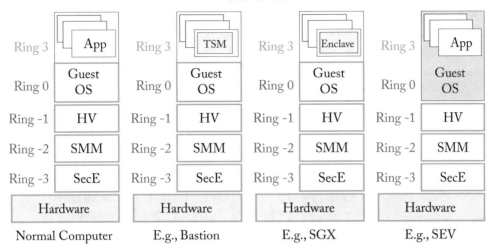

Figure 3.4: Example of how different architectures consider some of the privilege levels untrusted, and these are not in the TCB. Green outline shows which levels are trusted for these sample architectures. The three examples are Bastion [35], Intel's SGX [13], and AMD's SEV [12] architectures.

mentioned: software on software, software on hardware, hardware on software, and hardware on hardware. By introducing the different new levels, adding horizontal level separation, or designing the system to reduce the software TCB and consider some levels untrusted, different architecture for different threat models can be achieved. Especially, various combinations of trusted and untrusted levels in a system result in designs that protect from different software threats and attacks.

Typically, each processor architecture is composed of multiple execution privilege levels. At each privilege level, e.g., user level, there can be multiple software entities, e.g., there are many applications running on a typical workstation at one time, or there are numerous VMs (operating systems) running on a server. The processor architecture threat model has to specify which of the privilege levels are trusted and which are untrusted. Also, the threat model has to specify if entities at the same level are mutually trusting.

For example, today a reasonable assumption would be that the applications are mutually untrusting (one application may try to attack another) and that the operating system and hypervisor is untrusted, but the remaining levels are trusted (e.g., system management mode is trusted)—this would be a threat model of hardware architecture that supports trusted software modules, such as Intel SGX [13]. While typically there is a progression of levels (e.g., hypervisor then operating system then application), they are not necessarily more trusted in that same order, as the example shows.

A threat model has to say which of the levels and entities are trusted and which are not; see Figure 3.5. With the five levels there are $32 = 2^5$ possible combinations of security architectures based on different assumptions whether each level is trusted or not. So far the last two levels are almost always trusted (although they are not trustworthy, as there exist, e.g., SMM rootkits [237]). This leaves 8 possible security architectures, and a number of academic or commercial architectures have addressed the different combinations of threats, as shown on right-hand side of Figure 3.5.

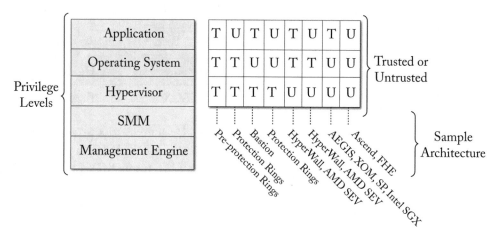

Figure 3.5: Given a set of privilege levels, each of these can be trusted or untrusted, depending on the secure processor design. The figure shows different combinations of trusted and untrusted levels and corresponding sample secure architecture that implements the assumptions about which parts are trusted and untrusted. The sample architectures listed are Bastion [35], Hyper-Wall [209], AMD SEV [12], AEGIS [205], XOM [136], SP [58, 129], Intel SGX [13] and Ascend [64]. While not a processor architecture, Fully Homomorphic Encryption (FHE) [65] is also included in this figure as it is one potential solution for processing data when all of the software is untrusted.

3.3.6 ARCHITECTURES FOR DIFFERENT HARDWARE THREATS

Modern computer system is composed of one or more separate physical chips. For example, memory is separate chip from the main processor. These chips are somehow interconnected, usually by wires or different types of busses on the motherboard that holds them all together. Because of this physical configuration, it may be easy to attack (or substitute) different components. Physical probing of wires, changing chips, and even modifying the physical chips form the different hardware attacks that secure processor architectures aim to mitigate where possible.

When designing a secure processor architecture, memory is often singled out as the component that should not be trusted as it is a passive component which can be easily removed and swapped with a different one. Physical probing inside the memory is typically not considered, but a memory chip can be always moved into another computer and its contents read out, e.g., the Coldboot attack [86]. I/O devices are also singled out as untrusted as they are typically from different, potentially untrusted manufacturing sources. Finally, the interconnect is simple to physically probe as well and should not be trusted. The upcoming sections provide details of hardware modifications that can help secure a system in light of all these components being untrusted.

Meanwhile, the processor chip is trusted. Reasons for trusting the processor chip is that chips are manufactured in increasingly more complex technologies (28 nm is standard today, with many companies fabricating chips as small as 10 nm or even smaller feature sizes). It is very expensive to actively probe such small features, whereas probing an interconnect bus which is orders of magnitude larger is easier. (Of course threat models develop over time, some recent works consider that different components inside the chip may be untrusted, as well as the interconnect inside the processor chip can be untrusted; see Section 7.3.)

An new trend of 3D integration presents a potential new approach to thinking about hardware attacks. The goal of 3D (also sometimes called 2.5D) integration is to stack multiple chips together in one package, or to have multiple packages stacked one on top of the other. For example, in a package-on-package configuration, a CPU chip is physically located below a memory chip. This can greatly reduce wire lengths and accelerate the data transfer between memory and CPU.

Because of the tight integration, it is now more difficult to probe the interconnect between CPU and memory, especially as attackers may have to physically disassemble the package-on-package configuration, which increases the difficulty of the attack. The 3D integrated systems may consider memory to be trusted again due to the higher difficulty of the attacks. On the other hand, tight integration of many components may lead to new types of attacks or side channels.

3.3.7   HARDWARE TCB AS CIRCUITS OR PROCESSORS

Key parts of the hardware TCB can be implemented as dedicated circuits or actually as firmware or other code running on dedicated processor, especially code running on dedicated processor seems to be preferred approach in industry. Intel Management Engine is reported to use and ARC processor or Intel Quark processor [169]. Meanwhile, AMD's Platform Security Processor is reported to be based on an ARM processor [135]. Effectively, there is a small processor running inside the main processor. Figure 3.6 shows the two options for realizing the security engine.

3.4   EXAMPLES OF SECURE PROCESSOR ARCHITECTURES

Numerous secure processor architectures have been proposed both in academia and industry. Interestingly, industry was probably the first explore secure architectures (or at least designed

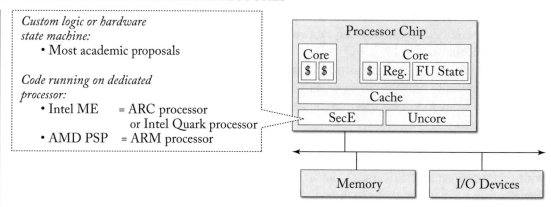

Figure 3.6: The hardware security engine, *SecE* in the figure, can be realized as dedicated circuits, or it can be implemented as a processor running some management code.

hardware features that now we associate with secure processors). Later, academia seems to have taken over, and in recent years it is again industry that is generating the most buzz about such architectures. Chapter 4 goes into detail on a number of these architectures, but they are listed here for reference and also some have been already shown in Figure 3.5.

3.4.1   ACADEMIC ARCHITECTURES

Starting in the late 1990s or early 2000s, academics showed an increased interest in secure processor architectures—XOM [136], AEGIS [205], Secret-Protecting (SP) [58, 129], Bastion [35], NoHype [112, 208], HyperWall [209], Ascend [64], CHERI [239], or Sanctum [43], to name a few. This is not an exhaustive list of architectures, but covers major academic architectures. First, researchers begun to be interested in protecting software applications (code) from hardware attacks and modification of the contents of the off-chip memories. Protections against operating systems were later introduced. Once hypervisors were introduced, they were co-opted to work with the hardware to provide the protections. As the hypervisor code begun to bloat, the hypervisor begun to be considered an untrusted entity and new protections were added to protect code and data against untrusted hypervisors as well. Finally, some architectures consider all software to be untrusted and explore how to perform computation on encrypted data. Most of the designs focus on single-processor systems. Multiprocessor security has focused mostly on the communication aspect (securing communication between multiple processors and memories), while individual processors in a secure multiprocessor system are often secured using ideas from single-processor designs.

3.4.2   COMMERCIAL ARCHITECTURES

Already in the 1970s, IBM included features such as logical partitions (LPAR) [79]. The security features were most common in main frame computers. Designers of later microcomputers did not incorporate many explicit security features, but in the late 2000s industry begun to again present designs such as Sony, Toshiba, IBM's Cell Broadband Engine [196] (with its security processor vault), Dallas Semiconductor's Secure Microprocessor Chip [146] (with its encrypted memory and self destruct mode), ARM's TrustZone [236], Intel's SGX [13, 95, 149], and AMD's SEV [50]. The commercial solutions leverage many of the academic ideas, but also their own pragmatic ideas needed to actually deploy the products.

One pragmatic feature, and potential weakness, of the architectures comes from use of dedicated security processors (inside the main processor). For example, Intel includes the Management Engine (ME) processor while AMD has a Platform Security Processor. These are used to realize some of the "hardware" features, such as managing the protections, or updating the page tables. These, often hidden, management processors run hidden code. While they are likely used to help quickly deploy the systems, they are vulnerable to software style bugs in the "hardware"—if the Management Engine code gets modified, for example, security of the whole processor is undermined [60]. This is, however, part of the tradeoff. Having all features in pure hardware is likely impractical due to time-to-market or cost constraints, so some "hardware" features have to be pushed to software. But that in turn leads to more software-style bugs that threaten the TCB of the system. Note, the choice of using dedicated circuits or an embedded processor is in addition to choice of use of microcode, which is widely used in processor design.

3.5   SECURE PROCESSOR ARCHITECTURE ASSUMPTIONS

Secure processor architecture designs typically involve a number of assumptions. At the end of the book, all the assumptions from all the book chapters are evaluated together.

3.5.1   TRUSTED PROCESSOR CHIP ASSUMPTION

The key to most secure processor architecture designs is the trusted processor chip assumption. It is assumed that the processor chip is the trust boundary for the hardware TCB. Everything in the processor chip is trusted, and everything outside is not trusted.

3.5.2   SMALL TCB ASSUMPTION

To prevent or minimize problems due to the TCB, the TCB should be small. It is further assumed that a smaller hardware and software TCB imply better security. The small TCB assumption is derived from two general ideas. First, less software code means it can be more likely audited, verified, and it will contain fewer bugs. Second, less hardware code likewise means it can be audited, verified, and will contain fewer bugs.

3.5.3   OPEN TCB ASSUMPTION

Using Kerckhoffs's Principle [167] from cryptography, the TCB should not contain any secrets, and it is assumed that the TCB can be inspected and analyzed. Especially, operation of the TCB should be publicly known and should have no hidden functionality. And, only secrets should be the cryptographic keys, to prevent security through obscurity.

3.6   LIMITATIONS OF SECURE ARCHITECTURES

Secure processor architectures are not a full solution to computer security problems, especially, secure processor architectures do not, usually, deal with the physical realization of the processors themselves. The general area of hardware security (as opposed to the architecture security) covers topics that are orthogonal to secure processor architecture design, but should be considered by architects when they think about their architectures. These are listed below.

3.6.1   PHYSICAL REALIZATION THREATS

Secure processor architectures assume that the manufactured chip, and especially the hardware, is correct. Research on hardware trojans [109], however, shows that malicious hardware can be inserted after the design time, e.g., at the foundry where the processors are manufactured. On the other hand, hardware trojan defense research shows how such trojans can be detected [212].

3.6.2   SUPPLY CHAIN THREATS

In addition to modification of the chip, which may in practice be very difficult, there are issues of the supply chain [178]. Modern servers, embedded system, etc., contain intellectual property (IP) from many designers and they are manufactured in variety of locations before being finally assembled into the finished product. At any of these stages in the supply chain a malicious component can be inserted into the system. For example, CPU can be correct, but the memory chip is malicious. Fingerprinting hardware modules and identifying correct ones, such as through use of PUFs, is one possible defense [55].

3.6.3   IP PROTECTION AND REVERSE ENGINEERING

Security should not be through obscurity, thus the design and hardware of the secure processor architectures should be known. Still, in a number of scenarios the designers may want to keep the hardware implementation a secret, e.g., due to fears of others stealing their intellectual property. Camouflaged logic [180], split-manufacturing [231], and other approaches can help prevent attackers from reverse engineering the design. Camouflaged logic aims to prevent one from deducing the circuit design, and behavior, by looking at the physical layout of the transistors. Meanwhile, split-manufacturing involves producing a processor chip in two or more foundries, where each one processes only few levels of the design. Split-manufacturing research shows how to divide the design into different parts (usually the design is split into back end of line, BEOL, and front end of line, FEOL) and each part is processed by a different foundry such that at a foundry cannot deduce final design based just on the parts it is processing.

A related topic to IP protection is threats of over-production. Ideas of hardware odometers [9] have been presented where new hardware features are used to ensure only legitimate devices can be authenticated. Furthermore, re-use or recycling of devices is dangerous as old, worn-out parts can be sold as new. Again, odometer features can give indication about the age of the integrated circuit chip and whether it is new, or has been used for extended period of time.

3.6.4   SIDE- AND COVERT-CHANNEL THREATS

Side- and covert-channel attacks can be used to leak the information based on the physical emanations (e.g., power, thermal, electro-magnetic). Such attacks can be damaging and used to leak sensitive information—typically the goal is to get the encryption key. Chapter 8 discusses side- and covert-channel threats, and protections, for processor architectures. A distinction needs to be made between information leaks due to the design of the logic, most often timing channels, and information leaks due to physical implementation. At the architecture level, the logic-related channels should be eliminated, e.g., due to processor caches. But then the physical implementation related channels may still remain, e.g., EM channels [6].

3.6.5   WHAT SECURE PROCESSOR ARCHITECTURES ARE NOT

Secure processor architectures are not hardware security modules (HSMs) such as IBM Crypto-Cards [101]. HSMs are dedicated, hardware modules that have extra physical security compared to typical processors (and secure processor architectures). HSMs may have tamper-resistant and tamper-evident coatings (e.g., the module may try to erase keys and shut down if it detects physical tampering, typically thanks to a wire mesh or other sensors that, when tampered with, signal an intrusion [14]). They further may be battery-backed, to ensure power-cycling of the whole system does not affect them (and that they have power to execute any proactive defenses in case of an attack where the rest of the system is shut down). Secure processor architectures mimic some HSM features (e.g., memory encryption) but usually do not deploy the more extreme measures (e.g., physical coatings).

A distinctive feature of secure processor architectures from HSMs is that they rely on platform features and modify only the architecture and digital logic. Secure processor architectures typically are not concerned with physical design, but aim to provide as much security as possible only through the architecture and digital logic level.

Secure processor architectures are also not security accelerators, such as dedicated devices for speeding up encryption or decryption. They almost always have dedicated hardware for acceleration of encryption, hashing, or public-key cryptography. But these features are used by the secure processor architectures' hardware to speed up the new protections it offers—a separate accelerator may still be present on a system if needed, e.g., for high-speed encryption or decryption of network traffic.

3.6.6   ALTERNATIVES TO HARDWARE-BASED PROTECTIONS: HOMOMORPHIC ENCRYPTION

A theoretical alternative to the myriad of hardware-based security modifications, or need to introduce new hardware architectures, may be Fully Homomorphic Encryption (FHE) [73]. In fully homomorphic encryption, operations are performed on the ciphertext and result in the creation of new ciphertext, which can later be decrypted to see the results of the computation. Importantly, the new ciphertext does not leak any information about the results, which can only be accessed by an entity with the proper decryption key. Currently, fully homomorphic encryption suffers from two practical limitations. First, the operations are very slow, making them impractical. Second, code is not protected as the protections only extend to the data.

Despite the limitations, with fully homomorphic encryption, the hardware manufacturer is not trusted. Actually, there is not a TCB anymore in some sense. All the inputs, intermediate data, and outputs are encrypted, so there is no plaintext information anywhere on the system that could leak out. Currently, there is an open research challenge between the two approaches. One one hand, if the hardware manufacturer is trusted, many operations can be done on a secure processor at high speed. On the other hand, any bugs or vulnerabilities could allow sensitive data to leak out, giving motivation of using cryptographic approaches such as FHE that do not depend on trusting the hardware nor the processor manufacturer.

CHAPTER 4

Trusted Execution Environments

This chapter first introduces Trusted Execution Environments (TEEs) and presents high-level description of the protections offered to TEEs by the Trusted Computer Base (TCB), and how the protections can be realized. It then presents a list of existing academic and commercial secure processor architectures, and the types of TEEs they offer as examples of possible design choices. It also presents TEE-related assumptions. The chapter closes by listing limitations of today's TCBs and the TEEs they create.

4.1 PROTECTING SOFTWARE WITHIN TRUSTED EXECUTION ENVIRONMENTS

TCB is the set of hardware and software that is responsible for creating the TEE environment. Software executing within a TEE is protected from a range of software and hardware attacks. The range of attacks that the software is protected from depends on the threat model of the particular secure processor architecture. The relationships between TCB and TEE are:

- TEE is created by a set of all the components in the TCB, both hardware and software

- TCB is trusted to correctly implement the protections; and

- vulnerability or successful attack on TCB nullifies the TEE protections.

Different secure processor architectures focus on protecting Trusted Software Modules (TSMs), also called Enclaves, while others on protecting VMs or containers. All of the code inside the TEE is given the same set of protections. There are no explicit protections that TEE gives to the different parts of the code in the TEE (except for the differentiation of the usual privilege levels, if the TEE contains a whole VM, for example). Users need to carefully consider what code runs within the TEE, especially if they use any external libraries or unverified code.

4.1.1 PROTECTIONS OFFERED BY THE TCB TO THE TEES

Confidentiality and integrity are the two main security properties that the TCB of a secure processor architecture aims to provide for a TEE. Confidentiality and integrity protection is from potential attacks by other software components or hardware components which are not in the TCB. There is usually no protection from attacks by malicious TCB or malicious code within a TEE. There can be multiple instances of the TEE running in parallel at the same time, e.g., multiple TSMs or Enclaves can be running on a system, each belonging to a different user. The TCB protects each TEE from potential attacks by the other TEE, e.g., there could be malicious code running in one instance of TEE trying to attack another TEE instance. Consequently, multiple mutually untrusting pieces of protected code can run on a system at the same time.

There is no uniform set of attack vectors that all the secure architectures aim to prevent. Most designs aim to protect from software-on-software attacks so that untrusted software is not able to attack the TEE or that mutually untrusting TEEs running on same system are not able to attack each each other. Note that for architectures that assume the OS or VM is untrusted, the TEE is protected from attacks the by more privileged OS.

Side channels are a weak point for many of the architectures, as they allow software-on-software side-channel attacks. Timing channels due to components in the processor are main reason for potential software-on-software side channels. Side channels can also be considered software-on-hardware attacks since the malicious software is abusing hardware behavior to learn information. Again, protections, if offered, are from attacks by untrusted components outside the TCB, or from attacks between mutually untrusting TEEs.

Hardware-on-software attacks are also partly considered by the different architectures. Especially, the main memory or system bus is often considered untrusted, so hardware probing to extract software secrets is often protected against. Hardware-on-hardware attacks are similarly often partly considered when concerning memory or peripheral devices, e.g., malicious network card connected to a memory bus should not be able to send malicious messages that would crash memory controller. Hardware attack protections, if offered, are from attacks by untrusted hardware components outside the TCB. TCB hardware (and software) is trusted by definition and attacks such as trojans are not considered (at the architecture level at least).

4.1.2   ENFORCING CONFIDENTIALITY THROUGH ENCRYPTION

Given the trusted processor chip assumption, and that everything outside of the processor chip is untrusted, symmetric key cryptography should be used to protect data going off chip to prevent hardware attacks. Figure 4.1 shows a diagram of a typical processor, with added encryption engine for protecting data going off chip. The security engine, or other part of the TCB, should encrypt data going out, and decrypt data going in to the chip. In the least, an encryption key for memory, K_m, is needed and needs to be stored securely in the hardware as part of the TCB. Chapter 6 deals in-depth with memory protection, and gives more details about how to implement the confidentiality protection through encryption.

4.1.3   ENFORCING CONFIDENTIALITY THROUGH ISOLATION

Protected software can be separated through isolation (controlling address translation and mapping) to prevent software attacks. Naturally, page tables are a well-known mechanism for con-

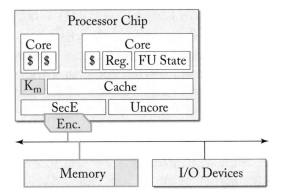

Figure 4.1: A diagram of a typical processor, with added encryption engine for protecting data going off chip.

trolling memory allocation. However, page tables are managed by the operating system or hypervisor, for second level page tables, either of which can be untrusted. In HyperWall architecture [209], for example, the hypervisor is untrusted to do the memory management, so HyperWall hardware manages the memory. The isolation must be managed by TCB and can be architected as a modification to the page table mechanisms (e.g., add another layer of translation or access control in addition to the OS or hypervisor managed page tables) or could be architected as dedicated memory management (hardware or software) module in the TCB that replaces the page table mechanisms.

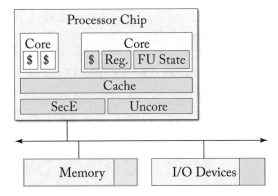

Figure 4.2: A diagram of a typical processor, showing in yellow the different components of the system that usually contain some state which need to be flushed between executions of different software.

4.1.4   ENFORCING CONFIDENTIALITY THROUGH STATE FLUSHING

State in the processor and elsewhere in the system needs to be flushed to ensure confidentiality from other entities that will later run on the system. Figure 4.2 shows the main parts of the processor that need their state to be flushed. While not often considered, any buffer or register in the processor can store data related to execution of a TEE. When the TEE is no longer executing (e.g., it was interrupted by another process, or it was terminated) all the relevant state needs to be flushed. Caches, registers, and memory are the obvious parts of the system that need flushing. However, there are also functional units and their state, e.g., the branch predictor state, or even state in the uncore or the security engine. Confidentiality leaks have been shown, for example, due to improper cleaning of the state, or "lazy" cleanup [152] that allowed sensitive data to be left in the processor state after TEE finishes execution. Chapter 8 covers side channels and gives more information about the state flushing, as lack of proper cleanup is one of the reasons for number of side channels or information leaks.

4.1.5   ENFORCING INTEGRITY THROUGH CRYPTOGRAPHIC HASHING

In addition to integrity protections, cryptographic hashing should be used to protect data going off chip to prevent hardware attacks and modification to the data. Figure 4.3 shows how the hashing is typically applied to data going in and out of the processor chip. All data going off chip (either due to explicit memory operations, or as part of saving and restoring processor state when TEE execution switches between different TEEs) needs to have its integrity protected. This can be achieved by having a hash module as part of the security engine, or other TCB component. Even if data is encrypted, someone could try to modify the behavior of the system by changing memory contents, even without decrypting the data. Most notably, an attacker could do a replay attack to replace current memory contents with previous good memory contents (which are now "old" but still would correctly decrypt). Having a hash value, that is fingerprint of the memory, stored in the TCB in the processor allows the processor to detect any changes. In addition, a nonce is needed. Chapter 6 deals in-depth with memory protection, and gives more details about how to implement the integrity protection through cryptographic hashing. Especially, raw hashing of memory is never used; instead hash trees are used due to their efficiency.

4.2   EXAMPLES OF ARCHITECTURES AND TEES

Various academic and commercial architectures have explored adding hardware support for securing the executing software at different granularities. The different secure processor architectures mainly focus on protecting TSMs or Enclaves, or whole VMs. In general, one key characteristic of the secure architectures is that there is some shared secret (key) between the TCB and the code running in the TEE, so that the hardware or software of the TCB knows how to decrypt and verify the code and data in the TEE.

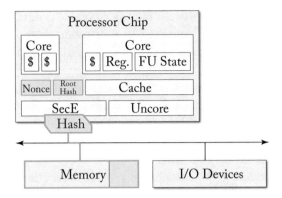

Figure 4.3: A diagram of a typical processor, showing off-chip memory contents being hashed using a hash engine; and need to store the reference hash value and nonce on chip.

Many of the architectures have concentrated on protecting discrete trusted software modules (TSMs), a term defined in [35, 58, 129]. More recently, these are also called Enclaves thanks to the popularity of recent SGX extensions introduced by Intel [149]. A mini-survey of the architectures is given below. First, the blow paragraphs focus on academic architectures, followed by discussion of commercial solutions that aim to protect TSMs or Enclaves. Second, architectures focusing on protecting whole VMs are presented.

4.2.1 ACADEMIC ARCHITECTURES FOR PROTECTING TSMS OR ENCLAVES

eXecute-Only Memory (XOM) [136] has user code stored in memory compartments such that one compartment cannot access another. XOM assumes that external memory is not trusted and, as a result, the data leaving the processor chip is encrypted. Also, each compartment has a session key associated with it and this key is used for encryption and decryption of the instruction and data stream. At any time only one compartment can be active, as its session key is loaded and used by the hardware for cryptographic protections.

AEGIS [205] is an architecture for tamper-evident and tamper-resistant processing. AEGIS assumes that all components external to the processor are untrusted. It uses encryption and hashing to protect sensitive data when it is outside of the processor core. Each application binary is encrypted and has an associated hash. The AEGIS hardware also has a private key, which can be used to sign a hash of the application so that a remote party can verify that the correct application was started.

Secret Protecting (SP) [58, 129] architecture proposed that software modules should have their confidentiality and integrity protected, when all the components except for the processor core are untrusted. SP defined the secure execution environment and introduced idea of protecting the software modules. Each SP processor possess a cryptographic key which can be used by the SP hardware to decrypt the software module's code and data. SP provides both user-mode [129] and authority-mode [58] variants. In user-mode there is only one owner of the device, while in the authority-mode there is a central authority that shares keys with a number of SP devices.

The Bastion [35] architecture extends the SP architecture by using a hypervisor to provide scalable trusted software execution and tailored attestation. Bastion protects TSMs with encryption and hashing, and multiple TSMs from different security domains can be active at the same time. Furthermore, in part due to the use of a trusted hypervisor, Bastion can provide flexible attestation of a selected set of TSMs.

Ascend [64] is an architecture that aims to protect from attacks by untrusted program. Although the program is trusted to perform the computations, however, from the perspective of the processor chip's inputs, outputs, and power consumption, an untrusted server cannot learn anything about private user data being processed. Ascend leverages ORAM to hide memory access patterns and differential-power analysis (DPA) resistance techniques to hide power consumption patterns [64].

Sanctum [43] aims to offer strong isolation of software modules running concurrently and sharing resources. It also protects against software attacks that infer private information from a program's memory access patterns. Sanctum architecture moves most of the TCB logic into the trusted software, reducing burden of need for complex hardware TCB.

4.2.2 COMMERCIAL ARCHITECTURES FOR PROTECTING TSMS OR ENCLAVES

Cell Broadband Engine with its processor vault [196] provides hardware isolated execution environments where code executing on a synergistic processing element (SPE) inside the Cell processor can be protected from the code executing in the rest of the system. SPEs are dedicated hardware processing units with associated and dedicated memory. The SPE and the memory can be locked out from the rest of the system and the code executing on the SPE can be isolated from even the main processing unit. The processor vault uses public-key cryptography to allow encrypted code to be sent to the SPE and decrypted, without being visible to the rest of the system.

ARM TrustZone [236] which creates two separate "worlds" for executing a trusted, secure operating system and untrusted, normal operating system. Even if the normal operating system or applications are compromised, the secure world is protected. In addition, memory and system buses can be tagged with an identifier of which world is currently executing, allowing some devices to be accessible only from the secure world, for example.

Intel's Secure Guard Extensions (SGX) [13, 95, 149] provides protections for trusted software modules, called enclaves, which are protected by hardware from untrusted operating system and other entities. Off-chip memory is protected through encryption. Although SGX did not intend to protect against side channels [104] like most of the prior academic architectures, researchers have found this as a potential weakness. Intel's new cache partitioning technology [140] can help avert some of the threats. It has some features of both Bastion or SP architectures presented in academia.

4.2.3 ACADEMIC AND COMMERCIAL ARCHITECTURES FOR PROTECTING WHOLE OSES OR VMS

NoHype [112, 208] introduced the idea of eliminating the hypervisor, while still allowing a number of virtual machines (VMs) to share the processor. Hardware was logically partitioned in NoHype such that each virtual machine obtained its resources on startup, and through the hardware separation mechanisms, one VM could not access resources of other machines. By eliminating the need for an active hypervisor, NoHype reduced the software TCB (at the cost of functionality, such as live migration).

HyperWall [209] proposed to protect whole virtual machines from the hypervisor (while still having hypervisor runing on the system to perform system management tasks). HyperWall introduced hardware for management of memory such that hypervisor could allocate and deallocate memory for virtual machines, but not observe their operation.

AMD's Secure Encrypted Virtualization (SEV) [50] is similar to SGX, but focuses on protection of whole VMs from an untrusted hypervisor. SEV ensures VM's memory is not accessible to other entities in the system. It has some features of HyperWall architecture presented in academia.

4.3 TCB AND TEE ASSUMPTIONS

TEE designs typically involve a number of assumptions. At the end of the book, all the assumptions from all the book chapters are evaluated together.

4.3.1 NO SIDE EFFECTS ASSUMPTION

Secure processor architectures assume no side effects are visible to the untrusted components whenever protected software is executing. Especially, the system is in some state before protected TEE software runs. Next, the protected software runs, often modifying the system and processor state. Finally, when protected software is interrupted or terminates, the state modifications are erased so that no information can leak to the untrusted components. Likewise, when multiple TEE run concurrently, one should not have visible side effects on another.

4.3.2 BUG-FREE PROTECTED SOFTWARE ASSUMPTION

The software (code and data) executing within TEE protections is assumed to be bug-free. The goal of any secure processor architectures is to create minimal TCB that realizes a TEE within which the protected software resides and executes. Software within the TEE cannot be protected from itself, e.g., if it is buggy or if some external libraries have been included that had vulnerabilities. Especially, code bloat endangers invalidating this assumption about benign protected software, as more and more code in the TEE increases chances of bugs. The TEE software could, however, try to attack other TEEs or the system. This, however, is a separate issue from bug-free TEE software.

4.3.3 TRUSTWORTHY TCB EXECUTION ASSUMPTION

Any vulnerabilities in the TCB can lead to attacks that nullify the security protections offered by the system to the TEE. Especially, problems in hardware state machines controlling the system could be exploited to nullify TEE protections. Or, problems in software or firmware running on the security engine, if it is realized as an embedded processor and not dedicated hardware circuits, can likewise be exploited to nullify TEE protections. Trustworthiness of the TEE then depends on the ability to monitor the TCB code (hardware and software) execution as the system runs. It is assumed monitoring of TCB mechanisms is able to: fingerprint and authenticate TCB code, monitor TCB execution and control flow, and protect TCB code (both software but also "hardware" if it is realized as an embedded security processor) through known mechanisms such as virtual memory, address space layout randomization, etc.

4.4 LIMITATIONS OF TCBS AND TEES

Hardware-protected execution environments are a great way to protect the critical code and computation. However, they come with some limitations and potential pitfalls for the designers and users. Each limitation can be seen as a research challenge: to find solutions on how to fix the limitations.

4.4.1 VULNERABILITIES IN THE TCB

Current designs allow for TCB-resident attacks, where the attacker uses the hardware protections to hide from the rest of the system. Examples of such TCB-resident attacks include SMM-based rootkits [59] where the attackers are able to subvert the SMM and use it as an attack vector. Because the SMM (ring -2) is more privileged than operating system or hypervisor, it may be impossible for system administrator to get rid of SMM rootkit once it is installed, especially, if there is no way to update or recover the SMM code. Another example are platform management engine-based attacks [214] which leverage the ME (ring -3) to hide and launch attacks. These can be even more damaging, as the ME is separate processor or microcontroller, and which cannot be updated or debugged, today. Once the TCB is compromised, any protections offered by the system are nullified and TEE is no longer secure.

4.4.2 OPAQUE TCB EXECUTION

Today there are often no means of auditing and accessing the code running as part of the TCB, especially code running the "hardware" that is actually implemented as an embedded processor, notably for the security engine. Proprietary code is usually a trade secret, with infrequent updates. Code signing, if deployed, further prevents users from themselves updating the code. This secrecy introduces type of security through obscurity, and there are well know attacks using the management engine as an attack vector to take over the whole system, e.g., ring -3 rootkits [59].

Code running and managing the TCB should be fingerprinted, and possibly authenticated. A hash over the TCB code can be computed at load time, by the hardware. Such hash could be available in a read-only register or memory location once the TEE code is loaded, but before the TEE code is executed. Attestation of the TEE could then include attestation of the state of the TCB.

Additionally, monitoring of execution of the management TCB code is an important feature so that correct execution of the system can be observed. There is a plethora of performance counters in todays processors. Yet, the most security crucial components, such as Intel's ME or AMD's PSP, do not seem to have any such features. Today, users simply have to trust the TCB code, which is often a black box with no run-time monitoring or reporting capabilities.

4.4.3   TEE-BASED ATTACKS

By design, trusted execution environments create a hardware-protected space wherein code can execute safely from outside inspection. This creates a number of challenges which, if not addressed, can allow for malicious code to leverage the TEEs as an attack vector, while the hardware features meant to protect TEEs help the attacker from being discovered, or stopped. Already, academics have presented potential attacks, such as the SGX-Bomb [107].

4.4.4   TEE CODE BLOAT

Code bloat (trusted hardware complexity or trusted software complexity) is a potential danger. As the TEEs are used to perform more and more functionality, there is more and more chance of a bug or vulnerability. Over time, researchers and programmers are finding clever ways to put more and more code into the TEEs. There exist Linux containers running inside Intel SGX [15] or a hypervisor running inside SMM [17]. The motivation for design of the TEEs is that they will protect a "small" piece of code which is bug-free (and ideally verified in some form). Meanwhile, this is often not true. Each new feature added for convenience or due to other needs of users endangers the platform as the TCB grows and TEE capabilities expand.

CHAPTER 5

Hardware Root of Trust

This chapter introduces the root of trust. It then discusses ideas of measurement and chain of trust. These ideas are used to demonstrate trusted and authenticated boot, remote attestation, and data sealing. The chapter also presents ideas regarding runtime attestation and continuous attestation. It next presents ideas for use of PUFs as root of trust. It also introduces ideas, and shortcomings, of using authentication for limiting what code can execute in the TCB or TEE. The chapter closes with a list of assumptions about the root of trust.

5.1 THE ROOT OF TRUST

The objective of secure processor designs is to provide authentication, confidentiality, and integrity protections for the TEEs they create. Thus, the system that realizes a secure processor architecture has to be trusted. The trust in most of these systems derives from two things. First, a root of trust, which is typically a cryptographic key or set of cryptographic keys. Second, trust in the manufacturer of the hardware and the hardware itself. This chapter focuses on the first aspect: the root of trust. To form the root of trust, cryptographic keys are needed for authentication, confidentiality, and integrity.

Authentication: The processor needs to authenticate itself, so a remote party or user can convince oneself that certain data or messages are indeed coming from a particular, trusted system. It is assumed that only one system (processor) exists that has a given (secret) key. Typically, public key cryptography is the choice for authentication: processor has a secret key, and a public key which is known to everybody. This key can be derived from a processor's secret key. Assuming manufacturer correctly initializes only one processor with an unique secret key, then any properly derived keys should be also unique.

Confidentiality: The processor needs encryption keys for confidentiality of data going out of the processor. For memory protection, ephemeral keys can be used, which are randomly re-generated on each system startup. For persistent data, the keys need to be generated in a systematic way. Typically, this combines in some way processor's secret key, and measurement of the system, as discussed shortly.

Integrity: The processor needs keys for integrity checking of data going outside of the processor. These can be derived as well from the secret key. One possible way is to use cryptographic hashes to generate all the keys from the root secret key. As long as the root key is unique and secret, the other keys should be as well. Getting access to a derived key, due to one-wayness of hashes, should not leak information about the processor's secret key.

5.1.1   ROOT OF TRUST AND THE PROCESSOR KEY

Security of the system is derived from a root of trust: a secret (cryptographic key) only accessible to the TCB components. From that secret key, it is possible to derive all the encryption, signing, and other keys. The root of trust key needs to be securely stored in processor chip the TCB, as shown in Figure 5.1.

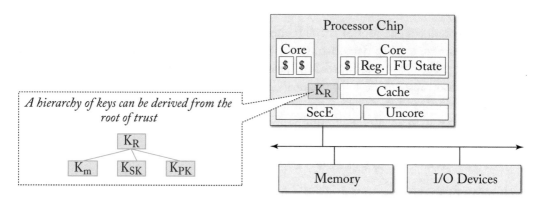

Figure 5.1: A diagram of a secure processor, showing its secret key, K_R, which is stored within the processor boundary. The secret key should only be accessible to the components in the TCB, and never leave the boundary of the chip. From the secret key, other keys such as for encryption or signing can be derived.

Each secure processor requires such a secret key, which should further be unique to each instance of the processor, i.e., each chip. One option is to burn it in at the factory by the manufacturer. However, this implies trust in the manufacturer and the supply chain, especially, it needs to be ensured the manufacturer does not hold a copy of the secret key after the processor is deployed. An alternative is to use PUFs. However, this requires addition of extra circuits that realize the PUF there is needed to ensure reliable operation of the PUF and error correction circuits are needed, and finally extra hardware module to derive cryptographic keys from the PUF output is needed.

5.1.2   PKI AND SECURE PROCESSORS

Public Key Infrastructure (PKI) is needed to distribute the certificates of the public keys of the processors. Especially, users need to know which are the keys belonging to the legitimate processors. The manufacturer, or a trusted third-party, needs to maintain the lists of known-good keys and the associated processors. Revocation is very difficult, and an open research topic: once a processor is compromised, its key can be extracted, and thus should be revoked. However, how should a manufacturer know that a particular processor is compromised, when the manufacturer likely never has access to the processor once it leaves the factory. There are also a multitude of integrators which actually integrate the chips into a working system, the secret key may be compromised there even before the product is sold. A potential solution here are PUFs: the manufacturer cannot set the key, and users can enroll the device when they receive it. Yet, between manufacturing and enrollment the processor or system could be modified or changed.

Typically, from the root of trust key, K_R, are derived keys for confidentiality, integrity, and authentication. A secret key, K_{SK} and corresponding public key, K_{PK}, are derived for encrypting data to be sent to the processor. This is data handled by the TCB, not the TEE as TEE software will have its own keys. A signature generation key, K_{SigK}, and a corresponding signature verification key, K_{VK}, are also derived. They are used to sign data generate by the TCB. Especially, the TCB's signing key can be used to sign user keys generated in the TEE, so that the TEE's keys can be tied, via the signature, to the underlying hardware that the TEE is executing on. Furthermore, for authentication, the signing key can be use as well.

As shown in Figure 5.2, the manufacturer (if the K_R key is burned in) can generate the public and verification keys, and their corresponding certificates. These can then be distributed

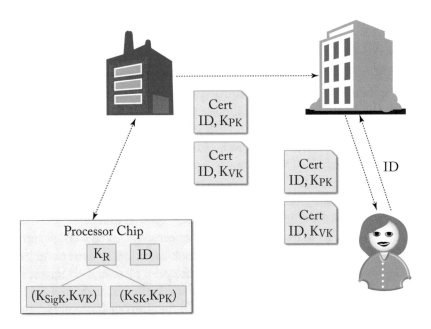

Figure 5.2: The manufacturer (top-left) initializes the secret key of a processor, and is able to derive different encryption (K_{SK}, K_{PK}) or signing keys (K_{SigK}, K_{VK}). The public (K_{PK}) and verification (K_{VK}) keys, and corresponding certificates, can be generated and given to the certificate authority (top-right). Given some ID of a processor, the users (bottom-right) can obtain the certificates for that processor's keys.

to certificate authorities, who in turn can communicate the certificates to users given some ID for the processor users' want to interact with.

The different keys are derived often by use of cryptographic hashes. Getting access to a derived key, such as the public encryption key, due to one-wayness of hashes, should not leak information about the processor's secret key. If the secure processor design leverages PUFs, it may require users to run their own key distribution solutions, or use existing, third-party PKI. Also, if the PUF-derived keys are never to be exposed, then the processor needs hardware or software that is part of the TCB to generate the derived keys and certificates, such that K_R truly never leaves the boundaries of the processor chip.

5.1.3   ACCESS TO THE ROOT OF TRUST

A physical attack or unscrupulous manufacturer or system integration could steal or leak secret keys. A more likely danger, however, is the code that is part of the TCB, and has direct access to the root of trust, which may leak the key unintentionally due to a bug, or due to some attack. Usually, BIOS or other low-level code is able to access the secret keys (e.g., to sign digital messages generated by the processor). This is usually code stored in flash memory, rather than a dedicated circuit. Bugs in the code can lead to exploits that can leak the secret key.

5.2   CHAIN OF TRUST AND MEASUREMENTS

When running TEEs on a system, it is crucial to know that it is correctly configured, e.g., only decrypt the file system if the correct operating system has been loaded. To achieve this, the trust begins with the secret key and a small piece of code which is implicitly trusted, such as a small BIOS. When a system turns on, this BIOS is the first code to execute. This code can then "measure" next piece of code that is to be loaded an executed, e.g., the firmware. Next, the firmware can "measure" the OS that starts up next, etc. Here, "measurement" is the act of generating a cryptographic hash of the code (and data) after it is loaded and before it is executed. The goal is to prevent someone from modifying this code (if the code is modified, the cryptographic hash will have a different value than the expected value).

Eventually, there is a set of hashes of the different components that booted up one after the other. Rather than to store each separately, each measurement can be used to "extend" the prior measurement. To "extend" means that hash value of the prior piece of code is concatenated with the next level of code and data when the hash is generated. Thus, at each level the hash depends on the code and data, and the hash of the prior level. In the end, one hash value depends on all of the prior code. This idea has been leveraged in the Trusted Platform Module (TPM) from which the term "extend" may have originated. The TPM [219] is a low-cost co-processor dedicated to security. The main objective of the TPM is to provide a root of trust in the platform. TPM helps to protect against software attacks by providing attestation and sealed storage mechanisms. The attestation, in particular, is based on a chain of cryptographic hash measurements of all the software layers.

When doing the measurements, different pieces of code can be included in the measurement. Usually, only the trusted code is to be measured. Meanwhile, there are different secure architectures which consider different software to be trusted. Figure 5.3 shows that when the measurements are done, different system levels may be included in the measurement. On the right side of the figure, two alternatives for how the measurements are done are shown. First, all the levels can be measured in succession: hardware measures the lowest software level, then each level measures the next level. The measurement is done via cryptographic hashing of the code running at that level. Second, some levels are "skipped." If a software level is not part of TCB or TEE it does not need to be measured.

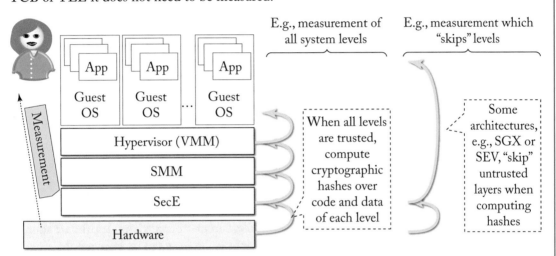

Figure 5.3: Left-hand side shows the levels in a typical secure processor architecture. Right-hand side shows two examples of different levels being measured. In the end, the user should receive the measurement, i.e., a cryptographic hash, that reflects the state of the system (the TCB and the TEE).

5.2.1   TRUSTED AND AUTHENTICATED BOOT

Trusted boot means that the system will only operate if all the measurements are correct, otherwise the system may not even start up. Authenticated boot means that the system does all the measurements, and users can access the measurement values, but the system will still continue to boot up and run even if the measurements are not as expected.

Currently, there is no standard action to be taken when the measurements are not correct. A system can stop executing or secret data can be prevented from being decrypted, as in trusted boot. But what if this system is part of critical infrastructure, e.g., an ambulance? Will the system refuse to run, e.g., ambulance's GPS unit will not turn on? Authenticated boot allows the system to run, but may not decrypt the hard drive data if the measurements are not correct. Here again, if this system is part of an ambulance, should life-saving patient data not be decrypted if the measurements are not correct? Failover modes where a standby module automatically takes over when the main system fails are not well explored in hardware secure processor architectures. Defining what to do when the measurement is not as expected is an open issue.

5.2.2 MEASUREMENT VALIDATION

The measurements taken during the trusted boot need to be validated. The main way to do the validation is to check the measurements against a list of known-good values. This, however, can be problematic. By definition of cryptographic hashes, even change in one bit of the input m will cause the resulting hash h to be totally different, compared to hash of the original input. Any update such as a new kernel version, or even addition of a new kernel module, or other update will cause all the hashes to change. Thus, it may become impractical to keep track of all the possible known-good values for all the possible permutations of system software. In the end, this problem may contribute to difficulty in deploying architectures in practice. A possible solution is to leverage the PKI used for key distribution, to distribute well-known hashes for variety of system configurations—or users can generate their own databases of known-good values based on software they or their companies use.

Figure 5.4a shows how a measurement is generated by the processor. It can be signed with the processors signing key. Then the user (b) is able to request a certificate from the trusted certificate authority. Given the certificate and the measurement, usually received over the network, he or she can validate the signature and know if the measurement came from legitimate source. Once the signature is validated, the actual measurement value needs to be checked. Mismatch in the measurement value means that some software in the TCB or TEE is not as expected.

5.2.3 REMOTE ATTESTATION

Remote attestation combines the root of trust with measurement, especially to attest that a system is running correctly, the measurements of the software as the system booted up can be signed with the private key of the processor, and sent to remote parties. These remote users can compare the known-good hash values with the received data and satisfy themselves that the system booted up in a correct configuration (or find that there is a problem). Because the hashes are signed with the private key of the processor, users are ensured that the measurement could not have been modified or forged after it was generated by the processor. It should be remembered that issues of replay attacks are critical: one needs to ensure at the protocol level that some correct (but old) measurement is not replied or sent again in future.

Figure 5.4c shows that the software (TEE) running on the processor can be attested using the measurements and signatures as well. The processor can make the measurement, of the TEE's code, sign it and send to the user. If he or she verifies the signature and the measurement, then he or she can send sensitive data to the TEE by encrypting it with a new session key. The session

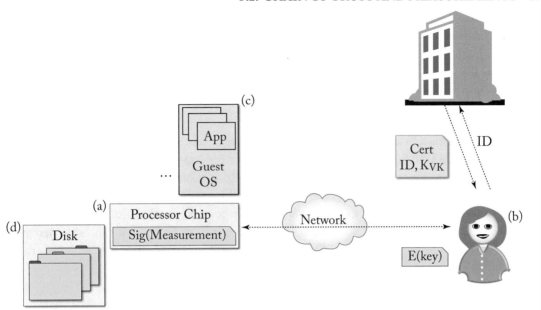

Figure 5.4: (a) A processor chip can generate measurements of the TCB (and TEE) and sign them with its signing key. (b) A remote use can obtain certificates from a trusted certificate authority, top-right, and use the certificates to validate the signatures of the measurements. (c) Once the measurement is validated, if it is for TEE, the user can send some sensitive information to the TEE, by sending encrypted information as well as sending the needed decryption key. The decryption key is protected by public key of the TCB or TEE. (d) The user can also send to TEE keys needed for decryption of data stored on remote disk.

key can be encrypted with the processor's public key, or more likely a key generated by the TEE, and signed by the TCB.

5.2.4 SEALING

Sealing is the act of encrypting some data with a key derived from a measurement. Conversely, unsealing, is the act of decrypting data with a key derived form a measurement. The idea of sealing and unsealing is to tie the data to a specific hardware and software configuration. For example, if hard drive data is sealed before system shutdown, then next time the system boots up, if there is any change to the measurements (e.g., any of the trusted software was somehow changed), then the proper decryption key cannot be derived and data cannot be decrypted. This prevents malicious uses from accessing data if they changed the software while the system was turned off (it does not protect against attacks while the system is running nor against time-of-check to time-of-use attacks discussed in the next subsection).

Figure 5.4d shows that the system can have some data stored on disk. Once the TEE is verified, the user can send back a cryptographic key (encrypted again with the public key of the TCB or public key generated by the TEE). That cryptographic key can be used to decrypt files and data stored on the disk.

5.2.5 TIME-OF-CHECK TO TIME-OF-USE ATTACKS

The main limitation of any measurement technique that measures (i.e., generates cryptographic hashes) the code is that it does not say anything about the code after it was measured. It only is based on raw measurements of static code as it was right before the piece of software executed. This leads to well-known class of Time-of-Check to Time-of-Use (TOCTOU) problems. The system may be correct at time t_0, but it may become compromised at time t_1, if user asks for the values of measurements at some later time t_2, he or she will get the hash value that states that system was okay at t_0, but no information about state at t_2 is gained. One solution to this problem is to perform continuous attestation at system runtime.

5.3 RUNTIME ATTESTATION AND CONTINUOUS MONITORING OF TCB AND TEES

Projects such as Copilot [168] have presented ideas about runtime attestation or checking of the state of the system. Most common approach is monitoring of the execution of the system. A co-processor can be added, where main processor executes normally, while a co-processor performs validation of the execution path. This can be also performed with performance monitoring counters now available with many processors.

Through observation of execution patterns, such as the number of branches taken, branch addresses, frequency of memory accesses, etc, a pattern of behavior of can be detected. This pattern can be learned and, at runtime, the performance counter data can be used to detect whether the program is behaving as expected or it has deviated. This can be applied to both the trusted and untrusted components, however, most work focuses on the untrusted components or the software that is being executed. Runtime attestation of the trusted components, i.e., the TCB, is so far not well explored.

A more structured approach is analysis of control-flow graphs. Source code can be analyzed to understand the control flow graph of software. Then, at runtime, again, the execution of a program can be checked to see if it has deviated from the expected control flow. This can be applied to the system and other software. At the hardware level, however there is often a state machine that dictates the behavior of the hardware—how to monitor the correct execution of such state machines is an open problem. Software can be monitored by hardware, but it is not clear how hardware, which is the lowest level in the system, can be monitored. A possible solution to this problem is to use hardware odometers [9]. The odometers work focuses on checking how long the hardware has been running (i.e., as an anti recycling feature), but similar counters or hardware features can check how hardware state machines are running.

5.3.1   LIMITATIONS OF CONTINUOUS MONITORING

Continuous monitoring can leverage performance counters and look for anomalies. This, however, requires knowing the correct and expected behavior of system and the software. The attacker can "hide in the noise" if they change the execution of the software slightly and do not affect performance counters significantly.

5.4   PUFS AND ROOT OF TRUST

PUFs leverage the unique behavior of a device due to manufacturing variations as a hardware-based fingerprint. A PUF instance is extremely difficult to replicate, even by the manufacturer and provides an unique identifier for the system. Thus, as mentioned earlier, PUF is very much like a secret key, and PUF response can be used in place of burning-in a key by the manufacturer. This can be used as a root of trust to establish whether software is executing on correct platform.

5.4.1   HARDWARE-SOFTWARE BINDING

So far, this chapter has focused on ensuring the correct measurement of the software, and checking at bootup time if software has been modified. But once the measurement is done, there is no real binding between the software and the hardware. To ensure that the software is indeed executing on a particular piece of hardware, hardware-software binding has been proposed. Hardware-software binding [81, 82, 123, 186, 187] is a method to precondition execution path of the software on the output of a PUF. Only on correct hardware will the PUF generate correct outputs that will cause software to execute as intended, if the software is moved to a different system, the PUF output by design will be different on another system, and the software will nolonger execute correctly. The hardware-software binding can be implemented via an indirection table in the software, where the PUF measurements are used to decrypt a table that controls the jump or call addresses used in the source code.

 To enable the hardware-software binding, at system runtime there needs to be a stream of output from the PUF to be used in the indirection table. A PUF can be read at start up time, and used as a key to a pseudo-random number generator. Storage of the PUF output may be challenging to secure, and in principle is not different from storing the secret key—thus hardware-software binding can be made to work with architectures that do not use PUFs, but do have a secret key, unique to each system. Or, a run-time accessible PUF can be used, such as the DRAM PUF [242] which can be queried at runtime. A run-time PUF reduces the need for storage of secret keys or generation of pseudorandom numbers from a stored key.

5.5   LIMITING EXECUTION TO ONLY AUTHORIZED CODE

Firmware (TCB) updates or protected software updates can be authenticated in the processor through use of signatures made by a trusted party. At manufacturing time, a certificate for the manufacturer's public and verification keys can be inserted into the processor. As shown in Figure 5.5, users can send code and data to an authorized party who has access to the signing key of the manufacturer. The authorized party can inspect the code, and if approved, generate a signature. The user can then send the code and the signature to the remote processor. The TCB components, with access to the certificate stored in the processor, can validate the signature, and, if correct, load, and execute the code.

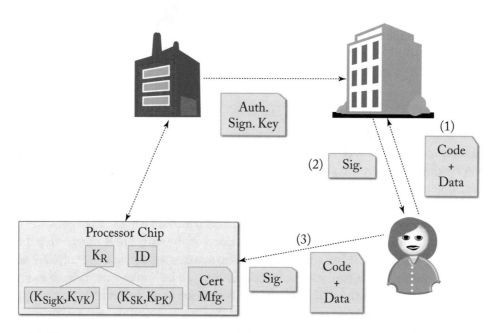

Figure 5.5: The processor chip can be embedded with a certificate for known-good verification keys, such that the processor TCB can authenticate messages, especially code and firmware updates. The manufacturer (top-left) can send signing keys to the trusted third party (top-right). Users can request the trusted third party, or even the manufacturer, to sign their code and data. The code and data can then (bottom) be sent to the processor. The processor can validate the signatures and only load the code if the signature is correct.

5.5.1   LOCK-IN AND PRIVACY CONCERNS

Privacy issue arise from any authentication mechanisms which limit what code can run on the TCB, or inside the TEEs. If only the manufacturer can update the device's TCB, then they are dependent upon for security updates and fixing bugs. Furthermore, requiring the manufacturer or another party to approve code that can run inside TEE further limits what code can be executed in the TEE. For example, as of time this book was written, to deploy Intel SGX Enclaves in production mode requires license from Intel and that Intel sign the code that is deployed inside any Enclave running on any server [105].

When a system is running and the TCB or TEE is being authenticated, use of signatures based on keys unique to a processor presents privacy issues. If a private key is used directly each time, one can know from which processor the messages are coming and a connection can be made between communication patterns and a specific physical machine. If a certificate authority is used for key and certificate distribution, it can know exactly when a specific processor is being used by a user who requests certificate about that processor. To solve such issues, Direct Anonymous Attestation (DAA), developed for TPM, offers some protections while allowing for remote authentication [29].

5.6   ROOT OF TRUST ASSUMPTIONS

There are typically a number of assumptions related to the Root of Trust. At the end of the book, all the assumptions from all the book chapters are evaluated together.

5.6.1   UNIQUE OF ROOT OF TRUST KEY ASSUMPTION

The unique secret processor key is assumed to be assigned to each processor so any two secure processors can be distinguished. The manufacturer is assumed to never assign two keys to different processors. If using PUFs, then the hardware needs to generate a unique key for each device even in event of errors or measurement noise in the PUF measurement from which the key is derived.

5.6.2   PROTECTED ROOT OF TRUST ASSUMPTION

The root of trust key is assumed to be protected and never disclosed. If keys are burned-in by the manufacturer, the manufacturer is assumed to keep secure database of the keys, and ideally delete the private keys once they have been generated. The manufacturer is entrusted with protecting the keys, and to never disclose those keys. If keys are derived from PUFs, then the enrolling party is trusted not to disclose the keys. The key is assumed to be stored in the processor chip and is part of the TCB and never disclosed. All TCB components (hardware or software) which have access to the key are trusted to never leak it.

5.6.3 FRESH MEASUREMENT ASSUMPTION

Authentication and data sealing give access to data to TEE software that is presumed to be correctly executing. Measurements are used to unseal data, and need to be fresh, meaning that they should be recently taken to avoid TOC-TOU attacks. Continuous authentication can be implemented to constantly check the TEE software. A mechanism needs to be in place to revoke access to sealed data if measurements change.

CHAPTER 6

Memory Protections

This chapter covers issues relating to protection of the main memory, typically DRAM, but also NVRAM in the near future. It overviews threats against main memory, dividing them into active and passive attacks. It then discusses mechanisms which can help ensure confidentiality, integrity, and access pattern protections for the main memory. It finally presents assumption relating to main memory protections.

6.1 THREATS AGAINST MAIN MEMORY

The main memory, today typically DRAM, is often a separate chip from the main processor chip. The main memory is connected to the processor chip via a communication bus that is a set of wires on the mother board. Following the secure processor chip assumption discussed before, both the main memory and the communication bus are typically assumed untrusted. As they are located outside of the processor chip, it is assumed that they can be more easily probed than any wires inside the processor chip.

Memory protection needs to consider both attacks from within the processor and from the outside of the system as the memory and the processor are separate physical chips. The penalty for these protections is the performance, e.g., encryption overhead, and area needed, e.g., to store root hash values in the processor.

6.1.1 SOURCES OF ATTACKS ON MEMORY

Memory is vulnerable to the different types of attacks listed as follows.

- Untrusted software running on the processor can attempt to access memory regions is is not allowed to.

- Malicious devices can attempt to use DMA to access memory regions they are not allowed to or snooping on the bus.

- Physical attacks on the memory bus can be used to extract data that is being communicated on the bus.

- Physical attacks on the memory itself, e.g., Rowhammer or Coldboot.

Especially, in addition to attacks such as probing the memory bus, DRAM chips can be easily removed and analyzed off-line. Moreover, with practical non-volatile memories

(NVRAM) on the horizon, data stored in the main memory will no longer disappear when memory is powered off, further requiring the contents to be protected.

6.1.2 PASSIVE ATTACKS

Attacks on memory can be divided into active and passive attacks. In passive attacks, the attacker passively eavesdrops on the communication, but does not try to alter any data. Passive attacker can try to learn the contents of messages—this can be prevented with use of encryption, i.e., encrypt all data before it leaves the chip boundary. He or she can also try to learn some information from the patterns of the memory accesses—this can be prevented with obfuscated memory accesses (usually using oblivious RAM techniques). The passive attacks focus on breaking confidentiality protections and passive attacks usually succeed if the designers did not correctly protect the information, such as encryption is not used while untrusted components have access to the sensitive data. Main passive attack type is listed below.

Eavesdropping Attacks focus on passively collecting information that is communicated to or from memory. However, getting access to data is not the only potential problem. Passively observing the memory access patterns can also reveal sensitive information. For example, access pattern to memory locations containing encryption S-Boxes can reveal which part of the S-Box is accessed, and can be used to leak information about the secret key. Encryption combined with hiding memory access patterns can protect against eavesdropping attacks.

6.1.3 ACTIVE ATTACKS

In addition to passive attacks, there are active attacks. In active attacks, the attacker can actively inject or manipulate the communication between the processor and the memory, or inject their own commands and request the memory to execute them. The active attacks focus on breaking either confidentiality or integrity—active attacks are usually successful if the designers do not properly authenticate communication, or if freshness is not ensured. Main active attack types are listed as follows.

Spoofing Attacks aim to inject memory data (or memory operations) without being detected. Attacker can inject memory reads or writes which will either change memory contents, or gain access to some memory. It may be possible, for example, to spoof memory read operation while the main processor is in sleep or low-power mode, and read the memory contents undetected. Reading memory contents possibly gives attacker direct access to secret data. Such attacks can be mitigated by proper authentication of memory access commands, or through use of encryption (where only legitimate entity has the right cryptographic key). Writing to memory can affect integrity of the system, for example the attacker can write new page table memory contents to attempt to modify the page table based isolation mechanisms. Such attacks can be prevented through use of hashing and message authentication codes.

Splicing Attacks, also called mix-and-match attacks, focus on combining data from two or more reads or writes to create a new, legitimate read or write. The splicing attacks, as the name implies, splice portions of different message, e.g., data payload of one message, with authentication header of another. If the whole read or write operation is not analyzed and authenticated correctly, attacker may be able to re-use portions of messages to create a valid memory operation. To protect against this type of attack, hashing and MAC can be used, however, the authenticated data needs to make sure to cover whole packet so that attacker cannot change part of message, while leaving the hash or MAC the same.

Replay Attacks, also called rewind attacks, leverage old messages, and send them again. If a nonce is not used, then there is no guarantee of freshness. Often, a monotonic counter can be used as a nonce to ensure that sender or receiver can discover that a message is old. Another, less common approach, is to have a global clock which can be trusted and used as a time reference for each memory operation. Replay protection involves both nonces (for freshness) and hashing or MACs to ensure that the correct nonce is always part of the message; otherwise a reply attack could combine part of one message, with a nonce from another message.

6.2   MAIN MEMORY PROTECTION MECHANISMS

As foreshadowed by the previous section, memory protection can be achieved by leveraging the three following techniques:

- confidentiality protection with encryption;

- integrity protection with hashing (typically using a hash-tree); and

- access pattern protection (typically leveraging Oblivious RAM techniques).

These are discussed in detail in the following sections. Confidentiality protection can be against attacks from different processor chips and I/O devices, or even against attacks from among different software running within a core (e.g., each trusted software module can have its memory encrypted and protected from other software, operating system or hypervisor). Likewise, integrity can be checked with respect to whole processor chip (one hash tree for each chip, as discussed below), or per trusted software module or per virtual machine. Access pattern protection requires special on-chip storage, to buffer and store some data locally, thus typically is done per processor chip. Although, per software module or per virtual machine protections could be designed in the future.

Figure 6.1 shows a simplified diagram of a multi-processor chip system, and depicts four, (a)–(d), major attack points. First, (a) is the untrusted software (either other user software or the management software) running on the processor. Memory protection is not primarily concerned with these—the focus is typically on the off-chip protection, which is done with encryption and hashing. Nevertheless, inside the processor chip the memory is decrypted so could be easily read by other software. The page table mechanisms can protect the data form other equally-privileged software (e.g., one application from another). To protect the data from higher-privileged software, data can be tagged so that the hardware (based on the tag associated with a piece of code

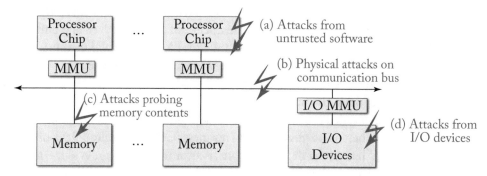

Figure 6.1: Simplified diagram of a multi-processor chip system with multiple memories and I/O devices all having access to the same interconnect. Four major attack points (a)–(d) are highlighted and explained in the text.

or data) can restrict access to the memory. Tagging memory requires extra bits to be included with each memory word, increasing size of the memory buses and storage elements. Second, (b) is the physical attacks on the interconnect. Physical probing outside of the processor chip's package is considered easier, and thus anything going off the processor chip and passing through the interconnect can be potentially snooped on or even modified, this is where the encryption and hashing is applied most often—but there is cost of encryption and decryption on each protected memory access. Also, data going off chip can be subject to access pattern analysis. Third, (c) memory contents can be probed or memory can be removed and analyzed off-line later. Well-known Coldboot attacks [86] showed that DRAM retention can be significantly extended by cooling down the DRAM chip. Moving forward, NVRAM which by design does not loose data when powered off, necessities protection of memory data as it can be recovered by anybody with access to the memory chip. Fourth, (d) I/O devices, typically connected to the PCIe bus and using DMA, can directly access memory.

Some of the threats can be mitigated thanks to memory management units (MMUs) and I/O MMU which serve to logically isolate the different processors, memories, and devices. However, the interconnect and memories can still be physically probed, or the I/O devices can directly access the contents of memories if I/O MMU is not configured correctly, necessitating encryption and hashing to protect memory contents. To overcome these limitations and come ahead of further potential threats, secure processor architectures typically add one or more of the three below types of protections.

6.2.1   CONFIDENTIALITY PROTECTION WITH ENCRYPTION

Memory encryption typically uses AES block cipher to transparently encrypt and decrypt data going to or from the DRAM, using an key re-generated on each system reboot. The key has to be stored on the processor chip, often in the MMU or in a special security management module. If same memory is to be shared among different processors they need access to the shared secret key. With multi-core processor chips, one MMU can be shared by the different processors, which actually removes need to have to explicitly share the key. In a multi-processor setup, the key has to be somehow shared by the processor chips. Sending of the key in plain text on the memory bus is not possible (attacker could easily snoop on the key). The solution can be public-key cryptography, however, this is very expensive in hardware and not often done. Most solutions today focus on having unique key(s) per chip without explicit sharing of the keys among different chips.

Memory sharing with devices, via DMA, is often needed. Here the device would need to have access to the key, or the data is sent in plain text from the device to the I/O MMU, which then can encrypt data going to the main memory. Partly sending data in plain text is not a good solution, design should encrypt data end-to-end. In practice, the devices doing DMA are I/O devices (network card, or a disk) and they would not do any encryption. Instead, the data they are processing is already encrypted (at some higher level in the software stack) so even though data going on the memory or PICe bus between device and memory is unencrypted per se.

To facilitate enabling and disabling encryption for certain memory ranges (e.g., memory shared with I/O devices is unencrypted), an extra bit in page table entries can be used by the OS, hypervisor, or the hardware to mark data needing protection. Data marked as not needing protection can simply bypass the encryption or decryption engine; see Figure 6.2. Data that does need protection will go through the engine—which add latency to the memory access. Counter-mode encryption can be used to speed up the process by pre-computing the encryption pads and then only doing one *xor* operation between data and the pad. This requires proper caching of the encryption counters used in counter-mode encryption to achieve good performance.

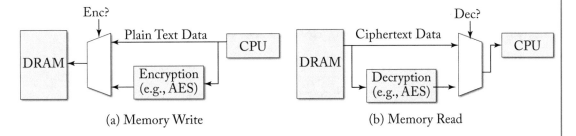

(a) Memory Write   (b) Memory Read

Figure 6.2: Block diagram of key components in encryption and decryption, shown not to scale. (a) For memory writes, select data can be encrypted, typically with a symmetric key cipher such as AES. (b) For memory reads, if the data is not decrypted, e.g., read by software not authorized, the CPU will only see the ciphertext.

In addition to using one key for whole memory encryption, different keys can be used for different applications or virtual machines, for example. The keys are managed typically by the MMU or a management engine—this may be the weakest link, as the key is accessible to the

MMU or management engine which can potentially decrypt all the memory traffic. However, assuming trusted hardware that manages the keys, it can switch which encryption key is used based on who makes the request. As long as the keys are generated by the hardware and only accessible to the hardware, the management software, e.g., hypervisor, can help manage which keys are in use, without actually getting access to the plaintext of the keys.

In addition to protection from attacks on the off-chip memory (DRAM), it may be required to protect from attacks from within the processor chip. For example, application may need protection from operating system, or operating system may need protection from a hypervisor. In these cases, encryption alone does not help and a tagging or ID-based approach is taken. Data can be tagged with a process ID or a virtual machine ID. When reading and writing data, the IDs can be compared between the requester and the data itself. For example, each virtual machine and the hypervisor are each associated with a separate memory encryption key. Outside the processor, memory is encrypted with the different keys, so each entity is protected from others. Inside the processor die, the hardware keeps track of the virtual machine IDs to prevent one entity from accessing another's code or data, while the code and data is in plain text.

In summary, secure architectures should use hardware to generate memory encryption keys on each boot cycle. Different keys can be used for different entities, to ensure data encrypted by one cannot be accessed by another. In addition to encryption, use of tagging or IDs is needed to control access to the data once it has been decrypted and brought into the processor.

Recently, commercial example of such solutions is AMD's Memory Encryption [50], which provides Secure Memory Encryption (SME) that allows for full encryption of DRAM memory contents. The processor is extended with hardware for encrypting and decrypting data going to and from the DRAM, similar to Figure 6.2. Furthermore, the memory encryption features in AMD are integrated with the AMD-V [5] extensions to create Secure Encrypted Virtualization (SEV) [12]. SEV form AMD tags all code and data with an ID of the virtual machine, called the address space ID, or ASID for short.

6.2.2   INTEGRITY PROTECTION WITH HASHING

Hashes are key cryptographic primitives used in integrity verification. Integrity protection leverages the fact that each input to a cryptographic hash will give a unique hash output (with negligible probability of two inputs giving same output, as determined by the size of the hash). Thus, the hash value uniquely represents the data, i.e., it is a fingerprint for the data. If the data is altered, the hash value will change: previously computed hash will be different from freshly computed hash and this difference indicates that someone changed the data. Rather than have to store a copy of the whole data, the relatively small hash value is sufficient for checking the integrity. Hashing is then the key part in integrity protection of memory contents.

Memory integrity protection most often focuses on the external attackers. The assumption is that the off-chip DRAM memory can be more easily manipulated or changed. Thus, after data is stored, and before it is used again, integrity checks are needed. Hashes require storage in a secure on-chip location so that there is a reference value which can be compared against.

A simplistic solution would be to hash the whole memory and keep one hash value. However, this is very inefficient as any small change to portion of memory requires hashing over the whole memory instead. A common alternative that is used are hash trees, also called Merkle trees [150]; a sample hash tree is shown in Figure 6.3a. Moreover, recall Figure 2.2 in Chapter 2 that presented hash tress in more depth.

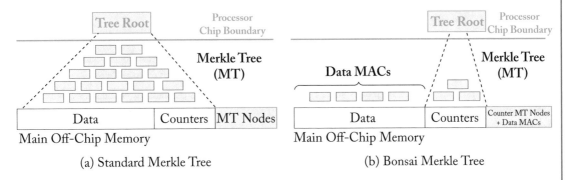

(a) Standard Merkle Tree   (b) Bonsai Merkle Tree

Figure 6.3: Comparison of (a) standard Merkle Tree used for memory protection and (b) Bonsai Merkel Tree which uses MACs and build memory tree only over the counters used in the MACs. The root hash is stored within the trusted processor boundary, shown in green.

Using Merkle trees for authentication [72] introduced the idea of building a tree of hashes, with the root stored in secure on-chip register. Advantage over the simple approach of hashing the whole memory is the performance of computing, and verifying, the hash values. After a piece of memory data is updated, only the hashes in the tree nodes on the path from the leaf node (that was update) to the root node need to be recomputed and updated. This will require only $\log_2(N)$ hash computations, where N is the depth of the tree. To verify a piece of data, again, rather than having to read whole memory and compute a hash, it is only required to read two adjacent memory locations, and compute the hashes on the way from the leaf node that is to be checked to the root node.

In the tree-based approach, the intermediate tree nodes (shown in darker gray in the Figure 6.3) need to be protected to prevent an adversary to compute, and insert, their own hash values. To address this problem, instead of using plain cryptographic hashes, secure processor designs use keyed hashes, also called MACs. The key needed to generate MACs is always stored in secure on-chip location. This way the tree nodes can be stored in the untrusted memory location. As a performance improvement, the tree nodes can be cached [72], as soon as node that is in the on-chip cache is encountered (and hashes match) verification can stop successfully.

Integrity protection is often combined with confidentiality protection, which itself often relies on encryption. With counter mode encryption [96, 177, 177, 194, 195, 204, 205, 244,

247, 254], memory is protected using a method where a counter is encrypted and the result is *xor*ed with the cipher text or plain text (for decryption or encryption). With counter mode, the encryption and decryption can happen in the background, while the critical path only requires addition of the *xor* operations. This, however, assumes that the right counters will be available at the right time (ideas of caching and prediction can be used to pre-compute and cache the encrypted counters). Counters can be public information, but their integrity needs protection; if the attacker can change counter value, he or she can change the result of encryption or decryption. Conveniently, the counters can be stored in the main memory, and protected by the memory integrity tree, as shown in Figure 6.3a.

Improvements on the standard hash trees are address-independent seed encryption and Bonsai Merkle trees [38, 176]. They aim to help with the performance issues introduced by the Merkle trees. The Bonsai Merkle trees are shown in Figure 6.3b. Address-independent seed encryption separates the encryption seed form the memory address. Using memory addresses, which were not meant to be used for security purposes, is not a good practice, and instead a logical page identifier can be used when encrypting the data (to prevent attacker from associating different data with different memory locations). With Bonsai trees, MACs are used for integrity checking of the main memory. An attacker cannot generate his or her MACs without access to the secret key. To prevent spoofing and splicing attacks, the memory location is included in the MAC. MAC can protect against spoofing and splicing attacks, but not for replay, thus a nonce (or a monotonic counter) is needed for each memory location. To protect the counter, see Figure 6.3b, the "Bonsai" Merkle tree can be created over the counters used for reply protection and logical memory location identifiers. The Merkle tree is much smaller than regular hash tree, which improves the performance. One drawback may be the MACs which are more expensive to compute and check. Architectures such as Intel SGX architecture [13], for example, use a tree of MACs provide data integrity and replay protection.

In addition to memory protection, often data from memory needs to be swapped out to disk (by operating system or hypervisor). Hardware-software secure architecture that support copying data to disk, may need to extend the integrity protection trees to cover disk as well. Bastion architecture [35], for example, has a memory integrity tree that has dedicated leaf nodes reserved for disk pages.

In summary, secure architectures should use hashing to check integrity of memory, and store the root of the hash tree in a secure on-chip location. They should use keyed hashes or MACs to ensure attackers are not able to generate their own hashes (to match their malicious data). For improved performance and to reduce the amount of data that needs to be integrity protected, approaches from Bonsai Merkle trees can be followed. Hash tress (using keyed hashes or MACs) combined with nonces can ensure protection of spoofing, splicing, and replay attacks.

6.2.3   ACCESS PATTERN PROTECTION

Besides getting access to raw data of the memory, observing memory access patterns can also reveal secrets. To address this issue, memory access pattern protection techniques have been proposed, most notably Oblivious Random Access Memory (ORAM) [74]. ORAM keeps the semantics of a program, but hides the exact memory access pattern of the program. Conceptually, ORAM prevents information leakage by maintaining all memory locations randomly shuffled: on each memory access, data read or written, and then reshuffled. The goal of ORAM is that any memory access pattern is indistinguishable from any other access pattern. An attacker seeing seemingly random access pattern does not know which data was actually accessed, and which one is a decoy access. While this clearly adds performance overhead, it can reduce the potential attacks.

ORAM can be realized by having a small, trusted memory on the processor chip, and the larger, untrusted, DRAM memory outside of the chip. For each memory access, the ORAM hardware or software makes a number of accesses to the external memory, with the goal of confusing the attackers. It heavily relies on the small, trusted memory that is located on-chip

Path ORAM [203] is one of the efficient ORAM implementation for secure processors. It is composed of two main components: a binary tree storage in off-chip main memory and an on-chip trusted ORAM controller. The binary tree stores the data content and is implemented using the main memory, DRAM. Example operation of Path ORAM is shown in Figure 6.4.

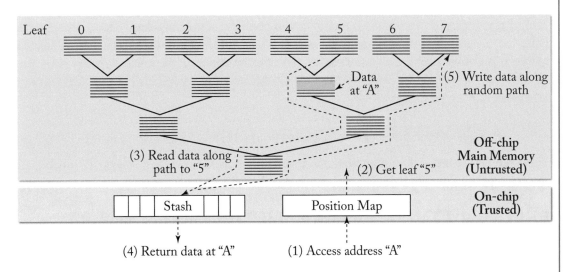

Figure 6.4: Example operation of Path ORAM showing data access and how all data blocks along a path are read into the stash, data is returned to the processor, and finally data blocks along different, random, path are written back.

Each node in the tree can hold up to N data blocks, and any unused space in a node is loaded with random dummy data blocks. The tree has a root node and number of leaf nodes, thus there is a path from the root to each leaf node, and on each path there are numerous data blocks (stored within the nodes of the tree). All nodes are encrypted and thus attacker cannot tell real data form random dummy data. ORAM controller is part of the trusted hardware that controls and manipulates the tree structure. The ORAM controller contains two main structures: a position map and a so-called stash. The position map is a type of a lookup table that can be used to map memory addresses of a data block to a path in the ORAM tree, i.e., which path in the tree should be traversed to find the needed data block. The stash is secure on-chip storage, or a cache, that can store a few data blocks. Data data stored in Path ORAM either needs to be stored in the DRAM (in the tree structure) or be located in the on-chip stash. To access data, first the path on which the needed data is located is looked up in the position map. Second, all the nodes (data blocks) on that path are fetched into the processor. The data blocks are decrypted and put into the stash; and meanwhile the requested data block is sent to the processor. Third, the requested address is added, via the position map, to a new random path in the three. Finally, nodes (data blocks) are encrypted and written back to the original path in the tree. Consequently, when next access is made to the same address, the new (random) path is accessed, hiding the true data access from the potential attacker.

ORAM ideas have been used in many settings, including in cloud computing [63]. As the memory hierarchy of a modern processor grows beyond just the DRAM and persistent storage, it is important to consider other elements of the memory hierarchy, such as the remote cloud storage, and how they need to be protected.

In summary, protection of the raw data of memory is not sufficient. The access pattern behavior needs to be protected as well, as it can leak information about what the protected software is doing. Amount of information that can be deduced from access patterns by the attacker can be reduced by shuffling the memory access patterns and hiding the true memory references among the random looking accesses.

6.3   MEMORY PROTECTIONS ASSUMPTION

There is typically one assumption related to the memory protections. At the end of the book, all the assumptions from all the book chapters are evaluated together.

6.3.1   ENCRYPTED, HASHED, AND OBLIVIOUS ACCESS MEMORY ASSUMPTION

Off-chip memory is untrusted and the contents is assumed to be protected from the snooping, spoofing, splicing, replay, and disturbance attacks. Encryption should be used to prevent snooping and spoofing attacks. Hashing should be used to prevent spoofing, splicing, replay (nonces must be used), and disturbance attacks. Meanwhile, oblivious access should be used to prevent snooping attacks.

CHAPTER 7

Multiprocessor and Many-Core Protections

The prior chapters of the book, like most secure processor architectures, have focused on uni-processor designs where there is only one processor. This chapter meanwhile discusses security issues in the design of secure processor architectures for computer systems which have more than one processor chip, or systems which have many processor cores on one chip. It first focuses on shared memory multiprocessors and distributed shared memory systems. It then covers security issues in many-core processor designs. The chapter ends with list of assumptions for multiprocessor and many-core secure architectures.

7.1 SECURITY CHALLENGES OF MULTIPROCESSORS AND MANY-CORE SYSTEMS

The overarching theme of the multi-processor designs is the communication. Uni-processor designs mainly focus on protection of the off-chip communication going to and from memory. The memory does not usually initiate communication so it is considered a passive element—it responds to requests from the processor. In multi-processor there is now a new dimension which is the processor to processor communication.

In multi-processor designs, securing inter-processor communication is a new challenge. In multi-processor designs, multiple processors, each on physically separate chip, are connected together, and connected to the memory. The interconnect is assumed to be more easily probed than chips themselves, thus motivating the need to secure the communication, and any data, that is transferred between different chips. Multi-processor secure designs need to ensure confidentiality, integrity, and authenticity of the messages going among the processor chips, and memory as well. Furthermore, one or more of the processors could be malicious (it is rather easy to swap out processors on the motherboard), thus some mechanisms are needed for authentication of the chips.

In many-core designs, the threats that multi-processor designs have worried about on the outside of the chip, have now moved into the processor chip. Many-core designs increase performance by including more and more processing cores within a processor chip (rather than connecting many chips together). Thanks to continued advances in manufacturing technology, tens or hundreds of processor cores can fit on a single chip today. Such expansion of cores within processor chip can, however, create a new threats: where one or more of the cores in untrusted. As in the multi-processor design, there are now many entities communicating, and the communication needs to be protected.

7.2   MULTIPROCESSOR SECURITY

Symmetric Multi Processing (SMP) and Distributed Share Memory (DSM) are two main design types for Multiprocessor systems. In SMP, there is typically a single memory bus which is connected to all the processors, and which is also connected to the main memory. All processors in SMP connect to the same main memory. In DSM, there is an interconnection network that connects all the processors. Also, as the name implies, the memory is distributed. In DSM, each processor has some memory that is physically near it, and thus is fast to access, and the rest of the memory is physically far (as it is near one of the other processors) and it is slow to access.

7.2.1   SMP AND DSM THREAT MODEL

SMP and DSM, also referred to as Non-Uniform Memory Access (NUMA), offer two ways of connecting many CPUs together. Figure 7.1 shows the two possible designs. In the figure, highlighted in green are trusted processor chips. From the perspective of a processor chip, all other processor chips, memories, or I/O components can be untrusted. Unlike with uni-processor designs, one danger of SMP or DSM designs is that one or more of the processors could be malicious. The different attacks can come from the untrusted software or hardware, especially, again, other processors can be potentially malicious.

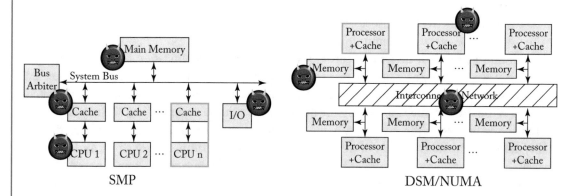

Figure 7.1: Block diagram of a typical Symmetric Memory Multiprocessor (SMP) system and a typical Distributed Share Memory (DSM). The usual threat model assumes the processor is trusted (highlighted in green) but the other components are untrusted, especially other processors can be sources of attacks.

7.2.2   SYMMETRIC MEMORY MULTIPROCESSOR SECURITY

In SMP designs, there is typically a single bus which is connected to the processors. The memory is also connected to this bus. All communication is done through the shared bus. The shared bus is typically part of the motherboard, and may be more easily probed than the processor chips themselves. Consequently, secure SMP designs need to protect the communication on the shared bus, plus potentially give protection from one or more of the other processors. They also need to protect memory as in uni-processor designs.

Confidentiality and Integrity of Communication. Work on secure SMP [195, 254] points to a number of solutions which can be leveraged to protect the communication among the different processors. Due to performance constraints, counter mode AES is the typical choice for encryption. The counter values can be pre-generated ahead of time, and only the resulting pads need to be xored with the data, adding minimal latency. The challenge is to keep track of the counters for the communication among the processors. One solution is to have each source-destination pair of processors use a dedicated set of counters for encryption of the data traffic. This has high storage overhead, but can achieve very good prediction rate (as the counters from each source to destination are used in sequential order). Storage can be reduced by having shared counters (one counter for each sender) or a global counter. Shared counters require less storage, but prediction rate is not as high. A global counter requires close synchronization of all the processors, something that multi-processor designs try to avoid.

For integrity protection, message authentication codes (MACs) can be used. Each sender can create a MAC for authentication and integrity checking of the message. As a performance improvement, AES in Galois/Counter Mode (GCM) [148] combines both encryption and authentication. Use of the so-called authenticated encryption mode of a symmetric cipher allows the system to generate the MAC at the same time as the message is being encrypted.

Confidentiality and integrity protections assume each source-destination pair of processors shares a secret key that can be used to do the encryption and MACs. If such a key is not pre-shared, public key cryptography would have to be used to allow two processors to agreed on the shared secret. In addition, a nonce (typically a monotonic counter) needs to be included in the MAC to prevent replay attacks where the attacker may try to re-send old messages.

Confidentiality and Integrity of Memory. SMP systems require memory protection just as uni-processor systems. Memory protection can be done using any of the variants of the Merkle trees discussed in previous chapter. As the memory is a shared resource, all the processors need to keep track of the memory updates, and possibly explicitly share the root node of the integrity tree—or individually update the tree root, but at all times they need to have a consensus about what is the correct tree root value. Especially, after each update, the tree nodes and the root node of the Merkle tree need to be updated so that all processors see the same value. An advantage of SMP systems is that each processor sees all the messages sent on the memory bus. Each processor can snoop on the memory bus and when it sees a message going to the memory it can authenticate the message. If correct, it can compute the new tree root value based on the message it has just seen.

One memory tree for whole system results in a performance overhead due to synchronization of the memory tree. An alternate solution is to maintain per-processor memory integrity trees. Each processor can maintain an integrity tree for the memory it is working with. Memory not being used can be excluded form the tree (certain tree leaf nodes are null) so that contents of this memory, presumably used by other processors, is not included in the integrity tree. Also, such memory locations excluded from a tree can be used for memory sharing. Data shared between processors can be put in physical memory that is not being checked for integrity (so after one processor updates it, the other can read it without encountering a validation error in its integrity tree). It can then copied from the shared memory to the memory being used exclusively by the processor, and which is protected by the integrity tree. When data is copied into the protected memory, the local processor's tree is updated and from that point forward the data has its integrity checked. When copying data from shared memory, the data needs to be authenticated, through use of keyed-hashes or digital signatures. Similar mechanism can be used for I/O devices that copy data to/from memory that is not protected.

Memory Access Pattern Protection. Because in simultaneous multithreading (SMT) every processor can observe all memory accesses, attacks based on traffic analysis are more easy (e.g., compared to uni-processor setup where the attacker has to physically probe the memory bus, here one of the SMP processors could simply be malicious and it is directly plugged into the bus). One counter-measure is the ORAM which can hide access patterns. Each processor in SMP would have to include an instance of ORAM to hide its access patterns from others. A disadvantage is that this may increase reads and writes on the memory bus, leading to more integrity related operations.

Key Management. Each processor in an SMP system needs to have its own key, from which keys for confidentiality and integrity checking can be derived. The key can either be burned in by the manufacturer, or PUF can be used to derive the key. The challenge is that the other processors in the SMP system need to be informed of the legitimate keys of all the other processors. How to efficiently implement the key sharing among SMP nodes is open problem.

The keys could be loaded into the processors at boot time by some trusted mechanism. For example, assuming there are no attacks at system boot up, the first processor to boot up can communicate its key to the others, and request their keys—or it can generate a key and send it to all the other processors so everybody uses the same key. This has obvious disadvantages that there may be attack at system boot up. Alternative is public key cryptography, but public key cryptography is too expensive in terms of computation to be added as a hardware block in a processor just for sharing keys among SMP processors. Nevertheless, in principle each processor could send the shared secrets to others using public key cryptography when the system is initialized, or when a new processor is added to the SMP system.

Storage of the shared secrets in each processor needs attention as well. An attacker could swap out a processor, try to extract the key, and then insert a malicious processor (but now with the right shared keys) into the socket. Thus, keys stored on the processors should be writeable (to initialize them), but only readable by the hardware (so there is no easy way to read out the key, short of invasive physical attack).

7.2.3 DISTRIBUTED SHARED MEMORY SECURITY

In DSM designs there is no longer a single bus that all processors can snoop on. Rather, processors may be connected in a mesh or other network. The connections are typically point-to-point, so all processor chips do not see all the traffic, rather data packets take different routes from source to destination. This allows many packets to be in flight at same time and improves performance compared to a shared bus in SMP. Furthermore, in NUMA the memory itself is also distributed, each processor is associated with a piece of memory that is physically co-located with that processor, as shown in Figure 7.1. Memory accesses are non-uniform, accessing local memory is faster than memory further away, allowing system to have lots of memory, with the benefit that data stored in memory close to the processors can be accessed faster. In addition, DSM systems have coherency protocols to keep track of which processor has the latest copy of which data. Thus, there are the coherency messages (a memory access on one processor may trigger a number of coherency messages needed to bring in latest copy of the data into that processor). From a security perspective, not all of this communication traffic is seen by everybody due to the point-to-point nature of the interconnect. This can be positive effect as traffic can be isolated or routed from untrusted components. But it can also be negative, as none of the processors have a total view of what is going on in the system.

Confidentiality and Integrity Protection of Communication and Memory. For processor-to-memory communication, confidentiality can be again ensured with a block cipher such as AES in counter mode, as it offers best performance (if the pads are pre-computed correctly and in time). All data's confidentiality could be protected with the symmetric key encryption. For integrity MACs can be used, or the AES GCM mode can combine both confidentiality and integrity; and Merkle tree can be used for checking the integrity of the main memory.

For processor-to-processor communication, encryption and MACs can be used. However, the plethora of cache coherence messages, makes it prohibitive to encrypt and hash all the messages to ensure security. One solution, proposed in a secure DSM design [177], is to only ensure that attacks that do not result in coherence protocol anomalies can are detectable. Because the cache coherency protocols have well defined states and transitions, it can be easily checked if a message comes which does not follow the protocol. Such a message can be dropped and trusted system software notified of discovery of a message that does not follow the protocol, which can further inform the users, for example. Only correct messages which are part of the protocol need to be authenticated. Only if a message cannot be authenticated, then an alarm is raised. With this approach, the number of messages needing authentication is reduced.

The communication in DSM is also quite fast. HyperTransport [40] is one example of interconnect used for DSM systems. It can operate today at up to 3.2 GHz clock frequency and with 4-byte links, with current standard. In such configuration, this requires processing 4 byte per cycle, while some of the best AES designs can do 0.25 bytes per cycle [8]. The performance of the interconnect is one of the key obstacles for security, performance should not be affected by the security, thus primitives such as AES have to be carefully integrated in the designs. This is also motivation for research and integration of light-weight ciphers that run faster.

Access Pattern Protection of Memory. In DSM all processors do not get to observe all the memory accesses or all the coherence messages. This can be leveraged to simplify memory access obfuscation. Routing of some packets could be controlled in a special way, or randomized, to hide the access patterns of one processor form the others. This may even remove the need for access pattern obfuscation all together.

Key Management. As in SMT, key distribution among the processors that are part of the DSM system is a challenge. Two main options are to pre-install the other processors' keys, or use public-key cryptography for processors to get each other's keys. Overhead of public key cryptography is usually prohibitive. Alternatively, and depending on the threat model, is to leverage the interconnect. One advantage of DSM is that there are point-to-point links, so two processors can communicate directly without others seeing the messages. This can be used to distribute pair-wise keys, where only the sender and receiver knows what the key is (assuming there are no external attackers). As in SMT, an attacker may be able to extract the keys, and he or she can potentially insert a malicious processor into one of the sockets and impersonate a correct one. Keys need to be protected from read out by any untrusted software (potentially even untrusted operating system or hypervisor).

7.2.4   SMP AND DSM TRADEOFFS

While SMT is becoming an older design style, with most processors moving to DSM, one benefit of SMT is that encryption and other cryptographic computations can be done fast enough in relation to (the slower) SMT data rates. This allows for encryption and integrity checks being done on reads and writes by each processor. Efficient key sharing among SMP processors is an open-issue today. While optimizations such as separate integrity trees can potentially improve performance of the system compared to a single tree.

Point-to-point links in DSM can be leveraged for security, as not all processors see all the communication and messages. They can also aid in sharing keys among the processors. Challenges of DSM include rather fast interconnect, necessitating either very fast cryptographic modules, or judicious protections of only certain messages passed between processors chips.

7.3 MANY-CORE PROCESSORS AND MULTI-PROCESSOR SYSTEM-ON-A-CHIP

Many-core processors and multi-processor system on a chip (MPSoC) designs share the characteristic of having many processing cores on the same chip. Similar to SMT or DSM, there are many processing elements, but instead of these processors being on separate chips, and the interconnect being exposed on the motherboard, they are all inside the same chip. In this configuration, the memory is typically a separate chip. However, with recent trends in 3D integration [71] the memory may also be in same package, or in a package-on-package (PoP) configuration.

Typically, many-core processors have same type of a processor core, while MPSoC combine various cores, often from different manufacturers. In both cases, the cores are connected by a network-on-a-chip (NoC). The are three main components of the NoC: the processors, the routers that are used to route traffic between processors, and the wires carrying the data. A sample figure of a multi-processor system-on-a-chip, along with some key components of the network-on-a-chip, is shown in Figure 7.2.

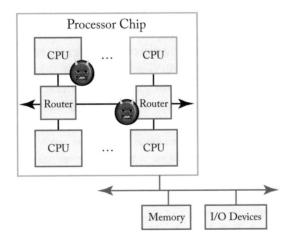

Figure 7.2: Potential threats agains many-core systems include untrusted processors, routers, or interconnect between the routers. Off-chip components such as the main memory or I/O devices are untrusted as well.

7.3.1 MANY-CORE AND MPSOC THREAT MODEL

Unlike SMT or DSM, designs that focus on many-cores and MPSoCs often assume that the processors (or CPUs), NoC routers, or the interconnect between the routers is untrusted, as shown in Figure 7.2. As with uni-processor and multiprocessor designs, the off-chip components (e.g., main memory or I/O devices) are still untrusted as well.

First, one of the processors or CPUs can be malicious, or contain a hardware trojan, e.g., [109]. The malicious processing element or the hardware trojan within it can drop outgoing or incoming packets, resulting in a denial of service. It can snoop on incoming packets—but only packets bound for that processor. It can alter outgoing or incoming packets as well. It can further inject new malicious packets into the network.

Second, the routers can be malicious or compromised. A router has access to all the packets that pass through it. It does not see all the packets in the NoC as not all packets pass through all routers. For the packets that do pass a particular router, it can drop packets, snoop on the packets, modify packet contents, re-order packets, or inject packets (including replaying old packets).

Finally, there are the wires on which the data is transmitted. As the wires are realized in the metal layers of the chip, they may be most easily probed, so a physical attacker can snoop on the packets. Note that physical attacks on the routers and processors are also possible, but they are subsumed by the hardware trojan assumption—hardware trojan can do as much if not more than a physical attack.

7.3.2   COMMUNICATION PROTECTION MECHANISMS

A data packet going from one processor to another is broken down into flits, and flits are further divided into phits. Each phit is akin to the smallest unit of data that can be transmitted. Flits are composed of many phits and are the basic unit that is moved around the NoC. Each flit may pass through multiple routers as it traverses the NoC, and each router is a hop in the communication network. As communication speed is critical, each flit should take one hop per cycle, thus the latency is the number of hops between the source and destination processor in the NoC.

Decryption of each flit at the 1 flit per cycle rate is not easily achieved. The performance criticality of the communication puts the biggest constraint on possible cryptographic protections. Optimized AES designs can take about 4 cycles per byte [8], while some lightweight ciphers can take about 1 cycle per byte [137]. Meanwhile, flit size is on the order of 16 bytes, while phits are about 2 bytes.

Two important solutions have been developed to secure NoC communication. First, since it is not the case that flits are generated each cycle, leveraging the average rate of flit injection into the NoC, combined with some prediction about the communication patterns and efficient cryptographic algorithm implementation can be one solution. Second, network coding approaches can be used instead of typical, strong cryptographic solutions. These two approaches should be analyzed with respect to the typical threat models for NoCs.

On top of these mechanisms for protection of the communication on the NoC, main memory protections can be built. As in SMT and DSM, memory encryption and integrity protections are needed to ensure external main memory is protected. Given that all the processors are on the same chip, it may be easier to have a single memory integrity tree, with the tree root value stored in some on-chip location accessible to every processor—but a synchronization mechanisms is needed to ensure only one processor only updates the value at a time. Given performance and synchronization issues, it may still be more beneficial to have separate integrity trees for each processor in the end.

7.3.3   3D INTEGRATION CONSIDERATIONS

There is a recent trend toward 3D integration (variably called either 3D or 2.5D). Especially with MPSoC systems, used in small embedded device, now commonly called Internet-of-Things, the main DRAM memory is integrated in the same package as the MPSoC, or in a package-on-package (PoP) configuration. This has notable performance benefit as the wires between processor chip and memory are shorter. Also, the vertical 3D integration can accommodate higher-bandwidth communication. Comparison between traditional design and 3D design is shown in Figure 7.3.

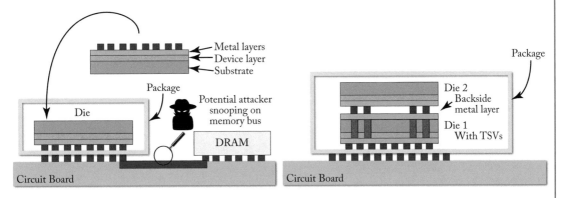

Figure 7.3: Example of traditional 2D design (left) and 3D integration (right). With the 2D design, processor and memory interconnect wires are exposed on the circuit board, allowing for potential attackers to snoop or invasively attack the. With the 3D integration, all wires going to or from memory are in the processor package.

The security advantage of the 3D or 2.5D integration is that the wires going between the processor and the memory are more difficult to access (they are either in the package, or sandwiched between two packages in the PoP configuration). This physical aspect of the design can be used to formulate a threat model where external attacks on processor-to-memory interconnect are not considered, removing the need to do encryption, hashing, or access pattern protection.

A related trend is the embedded DRAM (eDRAM). Embedded DRAM is dynamic random-access memory integrated on the same die as the main processor. It has the same potential benefit as 3D integration in that processor and memory connections are not exposed outside the processor package.

7.4 MULTIPROCESSOR AND MANY-CORE PROTECTIONS ASSUMPTION

There is typically one assumption related to multiprocessor and many-core protections. At the end of the book, all the assumptions from all the book chapters are evaluated together.

7.4.1 PROTECTED INTER-PROCESSOR COMMUNICATION ASSUMPTION

In addition to the existing uni-processor assumptions, designs with multiple processors or cores assume that the inter-processor communication will be protected. Confidentiality needs to be ensured such that if two processors are communicating, other processors cannot snoop on the communication. Integrity needs to be ensured such that one processor cannot modify memory of another processor. Communication pattern protection also needs to be ensured as malicious processor can observer the communication of other processors. Furthermore, the protected inter-processor communication assumption requires that different processors be able to mutually authenticate each other.

C H A P T E R   8

Side-Channel Threats and Protections

This chapter presents side channel threats, and protection approaches, for secure processor architectures. The chapter first introduces side and covert channels, and how they can lead to side- and covert-channel attacks. It then presents various processor features and how they contribute to information leaks. It also presents a simple classification of side and covert channels. It next discusses information leakage bandwidths due to various attacks. The chapter then discusses defense techniques, and especially highlights various secure cache designs. It also presents arguments for use of side channels as means to detect attacks, and it finally closes with discussion of side channel threats assumption made by secure processor architectures. Contents of this chapter are taken partly from this book's author's survey on the same subject of side and covert channels in processors [207].

8.1   SIDE AND COVERT CHANNELS

Side and covert channels have been introduced in Section 2.2.4. These channels are a means for communication of information. They are called side- and covert-channel attacks if the communication leads to some secret information leaking out. The attacks are often on confidentiality, i.e., the goal is to leak information out.

8.1.1   COVERT CHANNEL REVIEW

As a review, a covert channel is an intentional communication between a sender and a receiver via a medium not designed to be a communication channel. Covert channels typically leverage unusual methods for communication of information, never intended by the system's designers. They typically leverage observable changes in: timing of execution of some instructions or programs, power consumption of the system, thermal emanations of the system, electro-magnetic (EM) emanations generated as the system operates, or even acoustic emanations. Covert channels are easier to establish since both the sender and the receive are working together. Establishing a covert channel is a precursor to being able to establish a side channel (defined below).

Covert channels are a means for communicating information. A covert-channel attacks is an attack on a system's confidentiality, where the sender and receiver use a covert channel to leak out sensitive information. For example, a program inside an Intel SGX Enclave (i.e., the sender) may modify processor cache state based on some secret information it is processing, and then another program outside the Intel SGX Enclave (i.e., the receiver) may be able to observe the changed cache behavior and deduce what the sensitive information was, e.g., bits of an encryption key.

Figure 8.1 shows a schematic of a typical attack. First, (a) the sender runs some application that attempts to leak secret information to the receiver. Next, execution of the application or instructions can generate some emanations (2a) or cause processor state to be changed (2b); the sender and receiver know how the change in emanations or processor state is used to encode secret information that is to be transmitted. Finally, (c) the receiver is able to observe the emanations or change of the processor state (usually through timing changes), and deduce the secret information that was communicated by the sender.

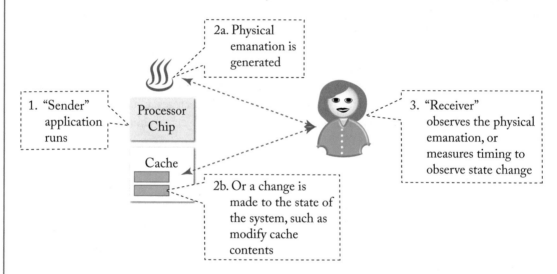

Figure 8.1: Schematic of a covert-channel attack. Side-channel attacks are similar, but the sender in the covert channel is an unsuspecting victim in the side channel.

8.1.2   SIDE CHANNEL REVIEW

In a side channel, the "sender" in an unsuspecting victim and the "receiver" is the attacker. A side channel is similar to a covert channel, but the sender does not intend to communicate information to the receiver, rather sending (i.e., leaking) of information is a side effect of the implementation and the way the computer hardware or software is used. Just as covert channels, side channels can be created using: timing, power, thermal emanations, EM emanations, or acoustic emanations.

The goal of a side-channel attack is to extract some information from the victim. Meanwhile, the victim does not observe any execution behavior change nor is aware that they are leaking information. This way confidentiality can be violated as data, such as secret encryption keys, is leaked out. Side-channel attacks work just as shown in Figure 8.1, but the "sender" from the figure in the victim and the "receiver" from the figure is the attacker.

Interestingly, side-channel attacks can work in "reverse." A side channel can also exist from attacker to victim. In a reversed attack, the attacker's behavior can "send" some information to the victim. The information, in the form of processor state change for example, affects how the victim executes, without the victim knowing there is a change. For example, the attacker can fill processor cache with data, causing the victim to run more slowly (a type of weak deinal-of-service attack). Or, the attacker can affect behavior of the branch predictor, causing the victim to execute extra instructions before processor is able to detect that these instructions should not be executed and nullifies their ISA-visible change [120].

8.1.3   SIDE AND COVERT CHANNELS IN PROCESSORS

One of the first mentions of what we now call side- or covert-channel attacks in reference to computers was made by Lampson in 1973 in his note on the confinement problem of programs [126]. Lampson discussed issues of confining a program during its execution so that it cannot transmit information to any other program, and he raised issues of side and covert channels. Since then, many researchers have explored side- and covert-channels attacks in depth.

From a processor architecture perspective, there is an intrinsic connection between the side and covert channels and the characteristics of the underlying hardware. Two key features need to be present for communication to exist. First, for communication to exist there has to be a channel. In processors, the channels are based on the shared hardware used by the different processes—because of spatial and temporal sharing of processor functional units among different programs, behavior of the hardware can be modulated and observed by the different programs. Second, for communication to exist there needs to be means of modulating the channel. In processors, many decades of processor architecture research have resulted in processor optimizations which create fast and slow execution paths, and which result in stateful functional units that are influenced by the instructions that execute on them. Taking the fast or slow path, or affecting state of functional units, which can be later observed, are some of the means of modulating information onto the processor hardware "communication channel."

The channels can be classified based on whether logical (software-only) or physical presence is needed. With software-only channels, the sender (attacker in covert channels and victim in side channels) and the receiver (attacker in both cases) are only utilizing microarchitectural features of the processor. Meanwhile with physical presence, the attacker can use physical means to probe the processor chip, requiring more resources and physical access to the target processor. Processor features leading to the different attacks are given in the next section, while the subsequent section classifies these attacks based on the logical and physical presence requirements.

As these types of attacks are well known, researchers have shown many defenses over the years. Nevertheless, almost all remain academic proposals. In particular, the designs which focus on eliminating the side and covert channels and their associated attacks often do so at the cost of performance, which is at odds with the desire to improve efficiently of modern processors that are used anywhere from smartphones, cloud computing data centers to high-performance supercomputers used for scientific research. There exists intrinsic interplay between the desire to prevent microarchitectural side and covert channels and the desire to further improve processor performance through microarchitectural enhancements. Any defenses need to consider the performance, power, and area overheads they impose on the whole computer system.

8.2   PROCESSOR FEATURES AND INFORMATION LEAKS

The sole act of executing an instruction and affecting one or more of processor's functional units' state can lead to a side or a covert channel. This is because there is an intrinsic relationship between processor microarchitectural features which allow today's processors to efficiently run various programs, but at the same time optimizations which lead to the side and covert channels. Because of the sharing of functional units among different programs, programs can in general observe, directly or indirectly, timing of the operation of the functional units. Knowing the design of the different functional units, timing in turns reveals whether the fast or slow execution path was taken. Finally, knowing one's own operations (i.e., the attacker program), or victim's program's operations, and the timing, an information leakage can be observed.

Figure 8.2 shows numerous sources of potential information leaks in a modern processor. Further, this book proposes that there are five characteristics of modern processors and their design which lead to microarchitectural-based information leaks that can form a basis for a side or covert channel.

1. **Variable Instruction Execution Timing**—Execution of different instructions takes a different amount of time, thus by observing how long some program takes to execute, some information can be learned about which instructions it is executing.

2. **Functional Unit Contention**—Sharing of hardware leads to contention, whether a program can use some hardware or not depends on the scheduling and what operations other programs are doing, leaking information about other programs trying to use the same functional unit.

3. **Stateful Functional Units**—Program behavior can affect the state of the functional units and output of the functional units or timing of instructions which depend on these functional units is related to state of the functional units, allowing for information leaks to occur.

4. **Memory Hierarchy**—Memory hierarchy is composed of many components designed to improve the average performance of a program, such as caches. However, memory accesses can be slow or fast depending on the state of the memory hierarchy, e.g., if data is in the cache or not, leading to many timing side-channel attack possibilities.

5. **Physical Emanations**—Execution of programs affects physical characteristics of the chip, such as thermal changes, which can be observed with on-chip sensors and lead to information leaks.

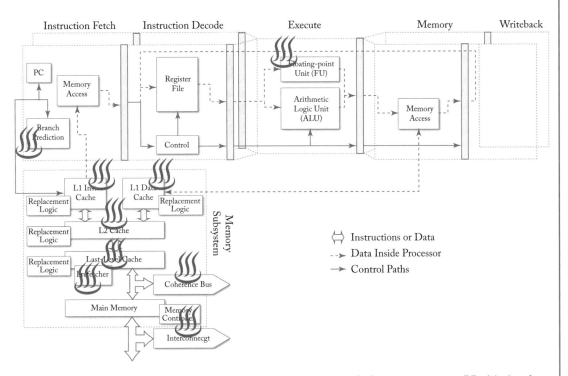

Figure 8.2: Diagram showing details of a typical multithreaded processor core. Highlighted in the diagram are main components which can contribute to side channels. However, any part of the processor which allows for sharing (spatial or temporal) of resources among different applications can potentially be a source of a side channel, even if no example attack using that component exists today.

8.2.1 VARIABLE INSTRUCTION EXECUTION TIMING

The job of the processor hardware is to perform different computations. Some computations are fundamentally simpler than others. Many logical operations (e.g., AND, OR, NAND, etc.) can be performed in a single processor cycle. Arithmetic operations such as addition also can be done quickly with use of parallel prefix adders or similar hardware circuits. Some other operations, however, such as floating point multiplication, do not have as efficient hardware implementations. Thus, processor designers have designed single- and multi-cycle instructions. As the names imply, a single-cycle instruction takes one processor cycle while a multi-cycle instruction takes many cycles to complete a task, usually due to the complexity of the instruction. Program timing will then depend on the instructions in that program. Thus, these fast and slow instructions lead to information leaks when computation is performed and the different timing of the instructions can be observed.

Eliminating the fast and slow paths would mean making all instructions take as long as the slowest instruction. However, performance implications are tremendous. The difference between logical operation and floating-point is on order of many cycles. Meanwhile, a memory operation (discussed in detail later) can take over hundreds of cycles if data has to be fetched from main memory vs. coming from a cache. There is an inverse relationship between the entropy among instruction execution timing and the performance. Greater performance differences between instructions implies higher entropy and more potential information leaks.

8.2.2 FUNCTIONAL UNIT CONTENTION

Processors are constrained in area and power budgets. This has led processor designers to choose to re-use and share certain processor units when having separate ones may not be beneficial on average. One example is hyper-threading, or simultaneous multithreading (SMT), where there are two or more execution pipelines per processor core. The multiple pipelines usually share the execution stage and the units therein. The motivation is that, on average, there is a mix of instructions and it is unlikely that programs executing in parallel, one on each pipeline, will need exactly the same functional units. Program A may do addition, while program B is doing memory access, in which case each executes almost as if they had all resources to themselves. A complication comes in when two of the programs attempt to perform same operation. If two programs, for example, try to perform floating point operations, one will be stalled until floating point unit is available. The contention can be reflected in the timing. If a program performs certain operation and it takes longer at certain time and shorter other times, then this implies that some other program is also using that same hardware functional unit, leaking information about what another program is doing. Thus, information leaks during computation when shared hardware leads to contention; this can result in attacker learning what types of computation, and when, the victim is performing.

Reductions in the contention can be addressed by duplicating the hardware. Today, there are many multi-core processors without SMT, where each processor core has all resources to itself. However, equally, many processors employ SMT and research results show that it gives large performance gains with small overhead in area. For example, SMT chip with two pipelines and shared execution stages is about 6% larger than a singe thread processor [31]. A two thread SMT is likely to remain in production for many years because of the evident benefits. Better performance/area ratios as explicit design goals imply at least some functional unit sharing will exist and in turn contention that leads to information leaks. The contention also becomes quite important in the memory subsystem, as discussed later in this chapter.

8.2.3 STATEFUL FUNCTIONAL UNITS

Many functional units inside the processor keep some history of past execution and use the information for prediction purposes. Instructions that are executed form the inputs to the functional units. The state is some function of the current and past inputs. And the output depends on the history. Thus, output of a stateful functional unit depends on past inputs. Observing the current output leaks information about the past computations in which that unit was involved.

A specific example can be given based on the branch predictor. The branch predictor is responsible for predicting which instructions should be executed next when a branch instruction is encountered. Since the processor pipeline is usually broken down into many stages, the branch instruction is not evaluated until much later after it is fetched. Thus, the hardware needs to guess which instruction to fetch while the branch is being evaluated, should it execute the next instruction, from the not taken path, or go to instructions from the taken path. Once the branch instruction if finally executed, the hardware can, if needed, nullify fetched instructions if it was found that there was as a mis-prediction and instructions from the wrong path were started to execute. Today, however, this is not always done. All ISA-visible state is erased, but microarchitectural state may remain modified, which can be detected via side-channels attacks.

To obtain good performance, branch predictor attempts to learn the branching behavior of programs. Its internal state is built using observation of past branches. Based on the addresses of the branch instructions it builds local and global histories of past branches. When a branch instruction is encountered, branch prediction is looked up based on the address of the branch instruction. If it was seen in the past, there is a taken or not taken prediction. Modern branch predictors can reach below 2 miss-predictions per 1,000 instructions [34]. Because of global history component of the branch predictors, different programs affect the branch predictor. A pathological program can "train" the branch predictor to mis-predict certain branches. Then, when another program executes, it may experience many mis-predictions leading to longer execution time and thus information leaks about which branches were executed.

Eliminating the branch predictor would negatively affect the performance of the processor and it is unlikely to be removed from modern processors. This is one example of how information leaks will exist as program's behavior (executed instructions) affect state of various functional units, that later affects others programs' timing.

8.2.4 MEMORY HIERARCHY

The processor memory hierarchy has some of the biggest impacts on the programs' performance, and information leaks. Not coincidentally, a vast number of research papers has focused on side and covert channels due to some functional unit in the memory hierarchy, some examples include: [220] [26] [93] [248] [172] [30] [127] [252] [217] [76] [155] [154] [28] [1] [3] [165] [222] [223].

Caches. The memory subsystem is composed of different layers of caches. Each L1 cache is located closest to the processor, it is smallest in size, but accessing data in the L1 cache takes about 1–2 processor cycles. There are separate L1 caches for instructions and data. The L2 cache is a larger cache, but at the cost of taking about 10 processor cycles to access data in the cache. The L2 cache can be per processor core or shared between multiple processor cores. L3 cache, also called Last Level Cache (LLC), is the biggest cache in size up to few MB, but accessing data in L3 takes tens of cycles. Finally, there is the main memory, sized in GBs, but requiring 100 cycles or more to access data.

Processor designers use the cache hierarchy to bring most recently and most frequently used data into the cache closest to the processor. This ensures that when there is memory access or instruction fetch, it can be fetched from one of the caches, rather than requiring going all the way to memory. Unfortunately, the fastest caches closest to the processor are also smallest, so there needs to be some policy of which data to keep in the cache. Often, the policy is some variant of the least recently used policy (LRU) that kicks out least recently used data or instructions and keeps most recently used ones. As programs execute on the processor and perform memory accesses, they cause processors to bring into the caches new data, and kick out least recently used data back to lower cache or eventually to the main memory (DRAM).

Keeping track of least recently used data in the whole cache is not practical, thus caches are broken down into sets, where each memory location can only be mapped into a specific set. Multiple memory addresses are mapped to a set. A cache typically has two or more ways, e.g., in a two-way set associative cache, there are two locations that data from specific set can be placed into. The LRU policy is kept for each set. For example, if memory addresses 0×0, 0×2 and 0×4 are accessed in that order, then when 0×4 is accessed, it will evict 0×0 from the two-way set associative cache as 0×0 was least recently used in that set.

Such design of the caches lends itself easily to contention and interference, which in turn leads to information leakage that is typically due to timing. The leakage can reveal whether some data is in the cache or not. Accessing data in L1 takes 1–2 cycles, while data in memory can take up to 100 cycles. Eliminating such leakages is difficult. Ideally, the cache replacement logic could search whole cache for the least recently used data, rather than just within a set. The Z-cache is one step toward that, however, its complexity has prevent it from being implemented [185]. Proposals for randomized caches have also been put forward [232]. However, currently if caches are removed, each memory access could take 100 or more cycles; it is not realistic to eliminate the caches form today's processors' memory hierarchy. Again, execution of different (memory) instructions takes different amount of time leading to potential for side or covert channels.

Prefetcher. Another component of the memory hierarchy is the prefetcher which is used in microprocessors to improve the execution speed of a program by speculatively brining in data or instructions into the caches. The goal of a processor cache prefetcher is to predict which memory locations will be accessed in near future and prefetch these locations into the cache. By predicting memory access patterns of applications the prefetcher brings in the needed data into the cache, so that when the application accesses the memory, it will already be in the cache or a stream buffer,[1] avoiding much slower access to the DRAM.

Hardware prefetchers attempts to automatically calculate what data and when to prefetch. The prefetchers usually work in chunks of size of the last level cache (LLC) blocks. Sequential prefetchers prefetch block $x + 1$ when block x is accessed. An improvement, which is most often used in today's processors, is the stride prefetcher which attempts to recognize sequential array accessed, e.g., block x, block $x + 20$, block $x + 40$, etc. [193].

Because hardware stride prefetcher fetches multiple blocks ahead, it will sometimes bring in data that the application is not going to use. However, depending on the physical memory allocation, that prefethed data may actually be used by another application. When the application accesses memory and measures timing, the blocks which were prefetched based on pattern detected for the other application will be accessible more quickly. In addition, if the exact details of the prefetcher algorithm are known, it is possible to trace back which addresses and how many addresses were accessed by the other application. Like for other parts of the memory hierarchy, prefetcher removal is not an easy option to defend against potential information leaks. Prefetcher removal would have largest penalty for applications with very regular memory accesses, but others will be negatively affected as well.

Memory Controller. The memory controller and the DRAM controller in the main memory are responsible for managing data going to and from the processor and the main memory. The memory controller contains queues for request from the processor (reads and writes, usually coming form the last level cache), it has to schedule these request, and arbiter between different caches making request to DRAM, which can cause contention.

The memory controller, which is a shared resource, becomes a point of contention. For example, two processor cores are each connected to the same memory controller and memory chip. Requests from each processor core need to be ordered and queued up for processing by the memory. Dynamically changing memory demand from one processor core will affect memory performance of the other core. While the memory controller attempts to achieve fairness, it is not always possible to balance out memory traffic from different cores. In particular, today's DRAM is typically divided up into pages and data from within DRAM is first brought into a row buffer before being actually processed (reads sent data back to the requesting processor from the buffer, or writes update it with incoming data). Many of today's chips use open-page policy that gives some preference to reads or writes to currently opened page (i.e., on in the row buffer). Memory accesses with lots of spatial locality may get preference as they hit in the open page—giving overall better performance as opening/closing new pages is expensive in terms of energy and time. Because of such optimization, again shared hardware leads to contention which in turn can be basis for side or covert channels.

---

[1]Some prefetchers place prefetched data in a dedicated stream buffer to limit cache pollution; stream buffer nevertheless is like a cache and accessing data in stream buffer is much faster than going to main memory.

Interconnect. Modern processors have replaced a typical bus that connected multiple processors and memory with more advance interconnects, such as Intel's Quick Path Interconnect (QPI) [102]. The interconnect is used to send data between processors and also for memory accesses in NUMA where main memory is divided up and separate DRAM chips and memory controllers are located near each processor. Such an arrangement gives each processor fast access to local memory, yet still large total system memory. However, timing of memory accesses can reveal information, such as accessing data in the DRAM chip close to the processor is faster than accessing remotely located DRAM at another core. In addition, locking and atomic operations can lock down the interconnect making memory accesses stall. Thus memory access timing can reveal state of the interconnect and leak information about what other processes are doing.

8.2.5   PHYSICAL EMANATIONS

While not a functional unit that performs some computation or data access, environmental sensors are becoming available on modern processor chips. One example is the thermal sensors. Modern Intel or ARM chips typically feature one or more sensors on each core, to help with thermal management of the processor, for example. The sensors can be read by the operating system and are generally exposed to user-space programs. Directly being able to read the sensor values opens up avenues for side or covert channels. For example, by executing a lot of instructions, one program will cause the chip to heat up. Meanwhile, another program can read the thermal sensor and thus learn about intensity of computation that the other program is doing.

8.3   SIDE AND COVERT CHANNEL CLASSIFICATION

In context of the processor, the information leaks can be turned into side or covert channels by attackers who are either running software on the same processor as the victim, or who are external to the processor. Figure 8.3 shows one potential classification of the attacks. The distinguishing features are the logical vs. physical presence, and whether the attacker needs to observe the victim, or just themselves.

Most of the microarchitectural attacks (and defenses) that are considered by the architects are at the left-hand side of the figure where software-only attacks are considered (i.e., logical presence). Architects do not usually control the exact physical features of the processor chip, such as a special package that may be added to protect from EM radiation leaks. They do, however, design and control the types of functional units, their sharing and behavior, thus have some control over potential information leaks. Nevertheless, many attacks in the right-hand side where physical presence by the attacker is required.

Because contention in the processor exists among different programs, attacker can observe the behavior of the victim, but also the victim can affect he behavior of the attacker. The top side of the figure shows that some attacks only require attacker to observe their own behavior (which is affected by the victim). Functional unit contention and stateful functional units are main sources of these attacks. The bottom side of the figure shows attacks leveraging observation of

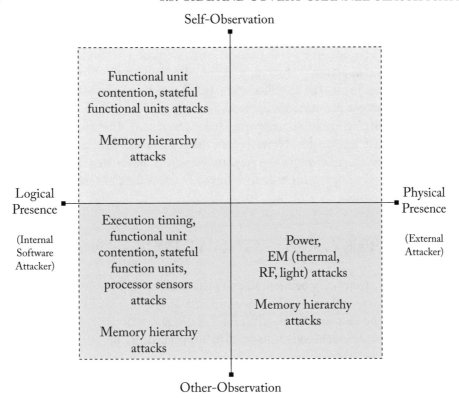

Figure 8.3: A classification of side- and covert-channel attacks based on whether physical presence is required, or attacker only needs to run some software on same processor as the victim; and whether attacker observes the victim ("other-observation"), or attacker only observes its own behavior ("self-observation"). Some combinations do not make sense, such as physical (external) attacker observing its own behavior, and are left blank.

behavior of the victim, where all five of the categories of processor features listed prior section contribute. The memory hierarchy is a contributor to attacks in most cases. Meanwhile, self-observation with physical presence, top-right of the figure, does not make sense, thus no attacks are listed there.

In terms of memory and cache attacks, in so-called internal timing attacks, the attacker measures its own execution time. Based on knowledge of what it (the attacker) is doing, e.g., which cache lines it accessed, and the timing of its own operations, the attacker can deduce information, e.g., which cache lines were being accessed by other applications on that processor. In the so-called external timing attacks, the attacker measures execution time of the victim, e.g., how long it takes to encrypt a piece of data; knowing the timing and what the victim is doing the attacker can deduce some information, e.g., was there addition or multiplication done during the encryption, potentially leaking bits of information about the encryption key. External timing channels often require many iterations to correlate timing information, however, basic principle is the same that attacker observes a victim's timing and the victim's timing depends on the operations it performs. The so-called trace-based attacks require tracking exact memory accesses, which is most easily done with physical presence.

Requirements of physical presence or only virtual presence (software-only attacks) impacts the types of attacks that are possible. Nevertheless, the boundary is fluid as introduction of new sensors onto processor chips, changes the requirements and makes new attacks possible with only software. Addition of new sensors contributes to functionality, but may create new avenue for side and covert channels.

8.4   ESTIMATES OF EXISTING ATTACK BANDWIDTHS

While new attacks are constantly being presented, this section aims to give an overview of past attacks and project future attack capabilities and what architects should expect. Most of the presented attacks in the past had bandwidths for information leakage in ranges of kilobits per second (Kbps) in optimal or idealized setting, and in bits per second (bps) or less in more practical settings. New attacks can reach Mbps.

The attack bandwidths have continued to increase over recent years. One of the first side-channel attacks was the 0.001 bps Bernstein's AES cache attack using L1 cache collisions [21]. Bonneau improved the attack to about 1 bps [28]. Around same time, Percival reported attacks with about 3200 Kbps using L1 cache-based covert channel, 800 Kbps using L2 cache-based covert channel, which reduce to few Kbps when they were done in a realistic setting [165]. Besides caches, a 1 bps proof-of-concept channel due to branch predictor [61] was presented. The work has since been updated [62] and latest results show about 120 Kbps channel. Further units inside the processor that have been exploited are the thermal sensors and recent work has shown 300 bps covert channel that leverages these on-chip sensors that can be used to measure physical properties of the chip (i.e., temperature) without physical presence [19].

In the realm of cloud computing, researchers have focused on virtualization and virtual machines. Ristenpart et al. showed 0.006 bps memory bus contention channel across VMs [174], and also 0.2 bps cross-VM L2 access-driven attack [174]. Xu et al. presented 262 bps L2 cache-based covert channel in a virtualized environment [243]. Zhang et al. show 0.02 bps L1 cache-based covert channel across VMs, using IPIs to force attacker VM to interrupt victim VM [253]. Wu et al. show 100 bps channel on Amazon's EC2 due to shared main-memory interface in symmetric multi-processors [241]. Hunger et al. show up to 624 Kbps channel when sender and receiver can have a well optimized and have aligned clock signals [100].

The presented attacks are only a subset of all attacks, and they only cover attacks without physical presence. Nevertheless, the bandwidths have increased over the years. These attacks leverage variety of functional units in the processor, and it should be expected that most processor features could potentially be used as source of information leak leading to side or covert channels.

Historical analysis shows that researchers are able to find side and covert channels based on variety of functional units requiring only virtual presence. Designers of new processor features need to expect that any additional features may be leveraged in a malicious way and should consider how the new addition impacts system security.

8.4.1 ATTACK BANDWIDTH ANALYSIS

The Trusted Computer System Evaluation Criteria (TCSEC), also commonly referred to as The Orange Book, sets the basic requirement for trusted computer systems [159]. The Orange Book specifies that a channel bandwidth exceeding a rate of 100 bps is a high bandwidth channel. Many attacks have long surpassed the boundary set by TCSEC. The cache-based attacks are highest in bandwidth as potential attackers are able to affect specific cache sets by executing memory accesses to particular addresses that map to the desired set. Other functional units let potential attackers affect the units state less directly. For example, many branch instructions are needed to re-train the branch predictor, which leads to lesser bandwidth channel.

Figure 8.4 shows the selected attacks and their bandwidths as a function of year in which they were presented. Bandwidth in bps on y-axis is plotted on log-scale for easier reading. It may seem unusual that as years progress, new published attacks do not necessarily have bandwidth better than prior attacks (due to expectation that new research should beat prior work). It should be noted, however, that some of the newer attacks are based on functional units that contribute less to the performance, so the bandwidth is less. The contributions of these attacks are clever ways of, for example, leveraging the branch predictor to leak information. Clearly, bandwidths are getting higher and have reached the bounds set by TCSEC for "high bandwidth channels," as shown by the red line in Figure 8.4.

When considering idealized attacks (shown as pentagrams in the figure) which are on the order of 100s Kbps, the "high bandwidth" boundary has long been passed in their case. These attacks, however, are usually specific to a single program, which often is the AES encryption algorithm. The attacks tend to be also synchronous, where the attacker and victim are executing synchronized (e.g., attacker and victim alternate execution on same processors). The synchronous attacks tend to have better bandwidth. For example, the L1 cache-based covert channel across VMs used IPIs to exactly force execution of the attacker to interleave with the victim. Dedicated attacker can thus come up with clever ways of improving bandwidth by having more synchronous attacks, approaching the idealized attack scenarios and further improving attack bandwidths. In summary, bandwidths for many side channel attacks today have already surpassed the bounds set by TCSEC for high bandwidth channels. Designers need to account for these potential attacks when considering new designs or analyzing existing ones.

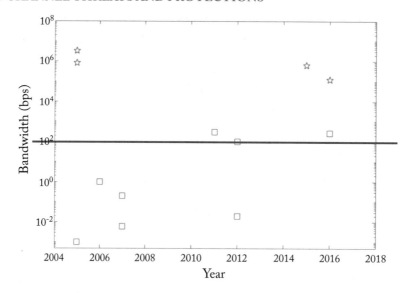

Figure 8.4: Scatter plot of bandwidths of select attacks, orders of magnitude bandwidth increase can be seen over last years for the non-idealized attacks (squares). Idealized attacks (pentagrams) show much greater bandwidths, but these are for specific cases, such as attacks on particular table lookup based AES software implementation or a particular RSA implementation. Today, some attacks are even better, reaching Mbps data rates.

8.5 DEFENDING SIDE AND COVERT CHANNELS

Since the side and covert channels depend both on the hardware, and the software that is running on that hardware, the defenses can be based on both hardware approaches and software approaches. The approaches are classified in Figure 8.5.

8.5.1 HARDWARE-BASED DEFENSES OVERVIEW

To mitigate side and covert channels, hardware architectural changes have been proposed including partitioning or time-multiplexing caches [163] [232], which have since been improved [124]. Such approaches essentially reserve a subset of the cache for the protected program. Other programs are not able to interfere with these reserved cache blocks. This prevents internal-timing, but external-timing attacks are still possible since measuring the protected program's timing from outside still can reveal some patterns about memory it is accessing. In addition, other applications cannot use the reserved cache blocks, effectively cutting down on cache size and the performance benefits it brings.

One of the best proposals focuses on new type of randomized caches [233]. Today's commodity hardware has caches where the mapping between memory and cache sets is fixed and

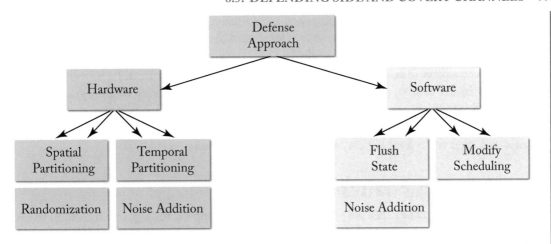

Figure 8.5: Hardware and software approaches to microarchitectural side and covert channel defense.

same for all applications. Randomized caches in effect change this mapping for each application. While not designed with security as a goal, the Z-cache [185] may have some of the similar properties where it searchers among many cache sets to find least recently used block for replacement, reducing applications' contention for a same set. In general, increasing associativity helps defend attacks, and number of caches aim to simulate highly associative cache without actually requiring implementation of a highly associative cache (which may have poor performance and use a lot of power).

Most recently, work has turned to on-chip networks and ensuring non-interference [234] or providing timing channel protection [230]. In [234] the authors re-design the interconnect to allow precise scheduling of the packets that flow across the interconnect. Data from each processor are grouped together and carried together in "waves" while strictly non-interfering with other data transfers. the authors in [230] observe, as this chapter does, that due to shared resources, applications affect each other's timing through interference and contention. The defense proposal is again to partition the network temporally and limit how much each processor can use the network so as to limit the interference.

Hardware-supported mechanisms have also been added for enforcing strong non-interference [218]. The authors propose "execution leases" which allow the system to temporally partition the processor's resources and lease them to an application so that others cannot interfere with the usage of these resources. This temporal partitioning again focuses on un-doing the original design where resources are shared at very fine granularity, and instead making it coarser, leaving to small potential leaks. The tradeoff is the performance impact of locking parts of processor for exclusive use of an application. The longer the application can keep the resources, the better leak protection, but also more negative impaction performance of other applications.

Processor architects have also proposed the addition of random noise to hardware counters [145]. The key to any timing attacks is to be able to obtain a timing reference, either within the application or somewhere from outside. In [145] the authors' approach is to limit the granularity and precision of timekeeping and performance counters mechanisms. By introducing more uncertainty in the timing, potential attackers are limited in their ability to get a good point of reference. Nevertheless, many applications are networked and can use external sources of timing, such as network time protocol [151]. Both inside and outside the computer system needs to be considered even when focusing on microarchitectural channels.

8.5.2   SECURE CACHE DESIGNS

To address the threats of cache side-channel attacks, focusing on timing, researchers have presented number of secure processor cache designs. There are at least 12 different secure cache designs presented in academic literature to date: [43, 56, 113, 129, 140, 141, 229, 232, 233, 245, 249, 250]. Most of the designs are only academic proposals, not yet implemented in real processors. However, CATalyst [140] is a proposal leverages Intel's CAT (Cache Allocation Technology) technology available today in Intel processors.

The *SP cache* [91, 129] uses partitioning techniques to statically partition the cache ways for separate use by the victim and the attacker processes, based on a process identifier. The *SecVerilog cache* [249, 250] also statically partitions cache blocks, but it distinguishes between low and high security levels. Instructions in the source code need to be tagged with the security level of each memory access instruction. The *SecDCP cache* [229] uses partitioning idea similar to the original SecVerilog cache, but the partitioning is dynamic; meanwhile, SecVerilog pre-set the size of the low and high partitions of the cache as system design time. The *NoMo cache* [56] uses an isolation approach by assigning each active thread; in a SMT (simultaneous multithreading), a number of cache blocks that are only accessible to that thread. The *SHARP cache* [245] uses both partitioning and randomization techniques to prevent victim's data from being evicted or flushed by other malicious processes. The *Sanctum cache* [43] uses page coloring [115, 211] approach to partition the cache, and a security monitor to ensure isolation of the secure data. The *CATalyst cache* [140] uses partitioning features from Intel's Cache Allocation Technology (CAT) [156] to partition the last level cache. The *PL cache* [232] provides isolation by partitioning cache based on cache blocks. The *RP cache* [232] uses randomization to de-correlate the memory address accessing and timing of the cache. The *Newcache cache* [142, 233] dynamically randomizes memory-to-cache mapping to prevent attacks. The *Random Fill cache* [141] also de-correlates the cache access timing by using random filling technique. The *Non Deterministic cache* [113] uses cache randomizes access delay to de-correlate the relation between cache block access and cache access timing.

Among the secure cache designs, there are two main features which they utilize: partitioning and randomization. Partitioning-based caches usually limit the victim and the attacker to only access cache blocks that are not accessible by the other. Randomization-based caches inherently de-correlate the relationship between memory address that was accessed by the victim, and the timing of the victim or the attacker. Randomization often implies either randomly bringing in data into the cache, randomly evicting data, or both. It could also imply randomizing the address to cache set mapping.

Partitioning-based secure caches are usually good at preventing external interference between attacker and victim, but are weak at preventing internal interference. Randomization-based secure caches are good at preventing external interference, and most of internal interference. Today, however, there is not a single secure cache that can fully defend all side-channel attacks based on timing.

8.5.3 SOFTWARE-BASED DEFENSES

Researchers have suggested clearing out leftover state in caches through frequent cache flushes [161]. This is clearly a performance-degrading technique, nevertheless it is able to prevent side and covert channels, as all application data and code is flushed from memory on context switch and when the application runs again it observes the long timing of main memory accesses. If the scheduling periods are long, application is able to fill up the cache and benefit from it. However, when scheduling periods are short, essentially the application will have to get all data from memory as the caches are flushed constantly. External timing attacks are prevented, and internal-timing can also be thwarted if the scheduling periods are short.

Outside of processor caches, to deal with the branch predictor based channels, clearing branch predictor on a context switch has been suggested [2] as well. Again, periodical clearing of the predictor state makes the current predictions not dependent on past inputs seen by the predictor, thus reducing information leak. However, such a defense is also a performance hit as branch predictors rely on learning the branching history of the running programs to give good branch predictor hit rate.

Addition of noise has also been suggested. For example, [97] explores reducing channels by introducing fuzzy time. In the work, a collection of techniques is introduced that reduce the bandwidths of covert timing channels by adding noise to all clock sources available to a process; this includes system time stamp counters and inputs from disk drives or network cards. Prior work has also introduced similar fuzzy time and created noise to defeat bus contention channel and to try to prevent attacks [77].

Other works focus on making the time and all events deterministic, such as in deterministic OSes designs [16, 240]. Rather than randomize the timing and add noise, all events are delivered at deterministic instants. Such approaches are not easily applied to caches or hardware features, but do help in virtualized environments where the hypervisor can control precisely delivery of packets or other timing information.

An example of pro-active attempt is the lattice scheduler which is a process scheduler that schedules applications using access class attributes to minimize potential contention channels [98]. In general, applications can be scheduled such that the contention is minimized. While existing schedulers do not do this, taking hardware into consideration a scheduler can run different processes on different cores, or time multiplex them such as to limit the contention. Of course, this assumes availability of many processes to be run and fairness of today's OS or hypervisor schedulers may be violated.

8.5.4 COMBINING DEFENSES OVERVIEW

Both software and hardware defenses can be used to try to mitigate side and covert channel attacks. Hardware can enforce spatial or temporal partitioning, create randomization or add noise. In software approaches, management software can modify (flush) processor state or change scheduling, it can also add noise. Other approaches are possible, but all approaches need to consider how added protections affect the system performance.

8.6 SIDE CHANNELS AS ATTACK DETECTORS

While side and covert channels are mostly considered as a negative aspect of a system, side channels can be actually used to detect or observe system operation. Measure timing, power, EM, and other behavior can be used to detect unusual system behavior and potential attacks. This approach is similar to using performance counters, but attacker does not know measurement is going on, or even how the measurement is being made. For example, existing research [49] explores how to use EM emanations to detect malware.

Consequently, a tension between side channels as attack vectors vs. as detection tools exists. Side channels are mostly used for attack today. But if the channels are someday fully eliminated, then their use as attack detectors will be eliminated. Thus, a middle ground needs to be found.

8.7 SIDE-CHANNEL THREATS ASSUMPTION

There is typically one assumption related to side channel threats in secure processors. At the end of the book, all the assumptions from all the book chapters are evaluated together.

8.7.1 SIDE-CHANNEL FREE TEE ASSUMPTION

The protected software assumes that the TEE is side-channel free. The TCB hardware and software should clean up processor state (both ISA-visible and microarchitectural) to remove any side channels. Memory hierarchy should especially defend protected software from side-channels attacks.

Despite protections from attacks by other software or hardware, TEE software still needs to defend against internal interference-based channels. In particular, software's own memory accesses interfere with each other and create timing differences that external attackers can observe when interacting with the TEE software.

C H A P T E R   9

Security Verification of Processor Architectures

This chapter forms a stand-alone chapter that is a mini survey of approaches for security verification of processor architectures. It first discusses motivation for the need for formal security verification. It then overviews the different levels (ISA, Microarchitecture, etc.) in a system and how verification can be performed at the different levels. It next discusses different security verification approaches. It then presents a mini survey of the different hardware-software security verification projects which exist today. The chapter closes presenting assumptions relating to the security vitrification of processor architectures. This chapter is based in part on this book's author's survey paper on the same topic [52].

9.1 MOTIVATION FOR FORMAL SECURITY VERIFICATION

News articles and opinion pieces by top security researchers constantly remind us that as computing becomes more pervasive, security vulnerabilities are more likely to translate into real-world problems [120, 138]. Computing systems today are very complex. If the design of the hardware, software, or the way the two interact is not perfect, then there may be security vulnerabilities that attackers can exploit.

Many secure processor architectures have been designed in academia to provide enhanced security features in hardware [36, 58, 108, 116, 129, 136, 205, 209, 239]. The proposed hardware-software system depends critically on the hardware and correct hardware-software interactions to provide the security. Yet, these secure architecture most often do not come with any formal specifications or proofs for security. In industry, increasing number of designs provide some hardware features for security, e.g., ARM TrustZone [221], Intel SGX [149], or AMD Memory Encryption [11]. These designs also all rely on the assumption that the hardware is correct—the industry designs also often do not have any publicly available formal security specifications nor proofs. With formal security verification, the designers could prove the system design and implementations are secure and trustworthy.

To help find potential security vulnerabilities and prove the designed system is trustworthy, formal methods can be used. Since the security of the systems depends on the correctness of the protections that both the hardware and software components provide, there is the need to verify the security of both the software and the hardware. Formal methods have been used extensively in the functional verification of hardware and software for a long time [20, 85, 114, 199]. Recently, the use of formal methods for verifying the security features of the hardware and the software levels of computing systems has emerged as an important research topic [52].

9.2   SECURITY VERIFICATION ACROSS DIFFERENT LEVELS OF ABSTRACTION

A hardware-software system is typically composed of multiple hardware and software levels, as shown in Figure 9.1. The typical software levels are: application, operating system (OS), and hypervisor. The typical hardware levels in a computer system are: ISA (Instruction Set Architecture), Microarchitecture, RTL (Register Transfer Level), Gate, and Physical.

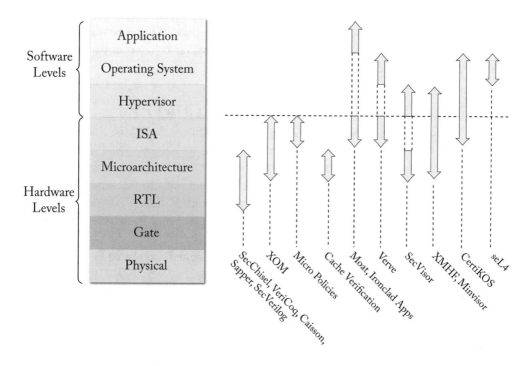

Figure 9.1: Hardware and software levels found in a typical commodity computing system are shown on the left. Sample academic projects and levels which they consider in the security verification are shown on the right. The sample projects are: SecChisel [53], VeriCoq [25], Caisson [134], Sapper [133], SecVerilog [250], XOM [136], Micro Policies [51], Cache Verification projects, e.g., [251], Moat [201], Ironclad Apps [89], Verve [246], SecVisor [191], XMHF [228], Minvisor [147], CertiKOS [80], and seL4 [118].

Traditionally, upper levels depend on the lower levels for functionality and security. A guest OS relies on the hypervisor to provide isolation from other malicious guests. If the more privileged hypervisor has a security vulnerability, the OS may not be able to provide any guarantees about security, as the hypervisor is more privileged in most architectures today. Or, a secure operating system cannot protect information leakage if the underlying hardware has a backdoor, for example. At the hardware level, for example, ISA is not secure if the microarchitecture that implements it has a bug; and a microarchitecture realized using a flawed RTL that implements it is likewise not secure, and so forth. The relationship is not strictly linear in that upper level always depends on all lower levels. Some of the secure processor architectures' have introduced hardware that allow higher software levels to be protected from intermediate software levels. For example, in Bastion [36] applications are able to communicate with the Hypervisor while bypassing the OS; or in HyperWall [209] a virtual machine does not need to rely on hypervisor for isolation as the hardware provides some of the basic memory management functionality. Thus, the security verification approach needs to consider which levels are important for security.

The software and hardware levels needed for ensuring security of the system constitute the trusted computing base (TCB) which contains all the software and hardware that need to be trusted (although may not be trustworthy). Thus, the TCB should be verified for security to make it truly trustworthy. Effectively, the TCB consists of different components in different levels, and security verification tools and methods should include all these levels when checking the security of the system. Figure 9.1 on the right side shows a sampling of existing academic projects and the different system levels that their security verification process covers.

Different projects consider different levels and some skip certain levels as well. For example, Moat [201] verifies applications with respect to ISA, and assumes hardware fully protects the applications from OS or Hypervisor, so OS can Hypervisor levels are skipped. When skipping certain levels, it needs to be ensure that they cannot affect the security properties offered by the system.

In summary, the system components at all the levels that make up the trusted computing base (TCB) should be verified for security to make sure the system is truly trustworthy. Security verification tools and methods should include all these levels when checking the security of the system. Meanwhile, some, usually lower, levels of the system are often assumed trustworthy and not checked; e.g, most architecture work does not consider the physical level and would assume that processors are manufactured correctly.

9.3   SECURITY VERIFICATION APPROACHES

The general flow of the security verification process is shown in Figure 9.2. The starting point of the verification is the actual system, either an already existing system or a design of some new system whose security properties need to be verified. From the actual system, or design, a representation of the system needs to be obtained, (a) in Figure 9.2. In parallel, the security properties of the system need to be specified, (b) in Figure 9.2. The security properties are closely tied to the system's assumed threat model. Security verification requires agreement on a threat model. If the threat model agreed upon does not represent the reality or the user needs, then the verification results may not be meaningful. For example, initially Intel SGX did not consider side-channel attacks due to processor caches [104], while attackers and researchers focused on this exact attack vector. Verification of Intel SGX, which would exclude side-channel attacks, would not have been as useful as verification that included consideration of such attacks.

Assuming the threat model is agreed upon, the security properties can be specified separately or together within the representation of the system, in which case (a) and (b) would be done together. The final step is the actual verification process which takes the system representation and security properties as input, and returns whether the verification passed or failed, (c) in Figure 9.2. If the verification fails, the design needs to be updated and re-evaluated, (d) in Figure 9.2.

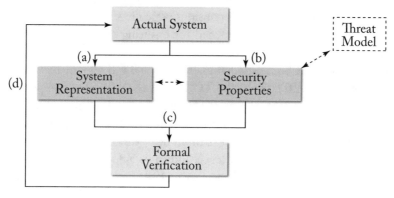

Figure 9.2: General procedure for security verification.

9.3.1   SYSTEM REPRESENTATION

In order to check if a system complies with some properties, a representation of the system that accurately expresses its behavior is needed. Ideally, the actual system description can be used, such as the hardware description language (HDL) source code for hardware components, or a programming language source code for software components. Otherwise, a model of the hardware behavior in the verification tool is needed. One reason a model may be needed is that the way a system is described in HDL or programming language may not be compatible with the verification tool that is being used, or the way the system is described is too complex for the verification process to handle. There is a danger, however, that the model may not accurately represent the actual system. If an automated method of creating a model (from source code representation) is not available, then model has to be created manually by engineers. When creating models manually, proving the correspondence between the model and the actual system is an open research problem.

Hardware components can be described with Hardware Description Languages (HDLs). The most popular HDLs are Verilog [215] and VHDL [139]. More high-level abstractions and reusability, than with traditional HDLs, is offered by hardware generation languages (HGLs) or hardware construction languages (HCLs), such as Chisel [18], Bluespec [157], or Genesis2 [192]. These languages usually focus on functional specification, but can be extended to include security related information, usually via annotations in the code. Examples of languages extended for security include SecChisel [54] or SecVerilog [250].

Software components can be described by their high-level implementation in programming languages such as C, C++, or Java. Like with hardware languages, system specification can be integrated with security-property specification, and include security verification related information inside the code itself as annotations. Examples include TAL [46] or Dafny [131].

Given the system representation, formal verification is done with respect that system representation. Most projects assume a trusted compiler or tool chain such that the system realization indeed matches the system representation, and does not contain extra hidden, or unwanted, functionality that may compromise the security of the system. For example, after verifying the C code of an application, there is still a concern that the compiler may not generate the correct machine code from the C code. A malevolent compiler might insert malicious code into the binary, as demonstrated in [216], where a virus-infected compiler as able to inject backdoors into applications during compilation. Researchers have developed trusted compilers that are guaranteed not to inject behavior that was not specified. One example of such a compiler is CompCert [132] which is a certified compiler that generates binaries from Coq code.

In summary, a system representation is needed to verify the system. Actual source code, or a model of the system, can serve as the representation. Even a verified system, however, is susceptible to vulnerabilities if the real system does not match the source code or the model—this may be due to attacks at manufacturing time or untrusted toolchains, for example.

9.3.2   SECURITY PROPERTIES

The main security properties that designers may be interested in are the confidentiality, integrity, or availability, listed in Section 2.3.1. Recall, confidentiality concerns prevention of the disclosure of sensitive information to unauthorized entities. If a system has some registers that should not be modified except by the hardware, then these registers require confidentiality protection form the system levels above the hardware. Integrity concerns prevention of unauthorized modification of sensitive information. If certain data should not be modified except by the operating system, then it requires integrity protection from the system levels above the operating system. Availability concerns ensuring provision of services to legitimate users when requested.

Note, as mentioned in Section 2.3.1, availability almost never can be achieved through use of one single secure processor design. For example, attackers can slowly use up memory or other resources of a system, making it impossible for protected code and data to execute in reasonable time. Availability, however, can be achieved by using many secure processor together, through redundancy, for example.

As was seen in Figure 9.1, typical system is broken down into many hardware and software layers. Often there is a linear relationship in that hardware or software components are protected from all levels above a certain level, which is least privileged level to be able to access (confidentiality) or modify (integrity) of that hardware or software component.

Confidentiality, integrity and availability of each system component that is part of the trusted computing base needs to be considered when specifying security properties for verification. Any unverified component could become a source of vulnerability.

9.3.3   FORMAL VERIFICATION

Typical formal verification mechanism include theorem provers or model checking. When using theorem proving approach, the security properties can be represented in terms of logical formulae. A logical formula serves as a limitation on the states the system is allowed throughout its execution. When using model checking, security properties can be expressed as invariants within a system, and their validity is checked against all possible execution paths of the system.

Theorem provers, also called proof assistants, use formal proofs to verify that a system complies with some given properties. Theorem provers aid the verification process by providing frameworks for creating a mathematical model of the system, for specifying the security properties, and for formally proving whether the model complies with the properties or not. Theorem provers are generally composed of a language (such as Coq [99]), and an environment for describing the proofs (such as CoqIDE [42]). There is a number of proof assistants used actively in academia and industry such as: Coq [99], Isabelle/HOL [158], PVS (Prototype Verification System) [162], ACL2 (A Computational Logic for Applicative Common Lisp) [111], and Twelf (LF) [88]. Theorem proving typically requires a lot of effort and time to complete, and learning the required tools is seen as one of the difficult aspects of verification using this method.

Model checkers, on the other hand, typically use an algorithmic search, which is performed over a system's representation and its states, rather than using deduction. Internally, model checker often rely on satisfiability modulo theories (SMT) solvers to analyze the state space. In general, checking whether a system complies with a given specification is an undecidable problem [41]. With model checkers, the problem is converted into a search problem with a reasonable coverage of input instances, however, it may not always give a solution. Especially, model checking has a well-known state explosion problem, which is the exponential growth of the states of the system that should be evaluated. State explosion leads to extremely long run times or exhaustion of the compute resources, thus preventing it from giving a solution sometimes. For fairly complex systems, model checking needs to use more abstraction and simplified models. However, as the level of abstraction gets higher, there is the risk of missing some important details of the system being verified.

The security property that is being verified using model checking has to be defined using a logical form. The checks can be done either for each transition or each state using invariants, pre- and post-conditions. The output can be positive (property satisfied), negative (property violated), or the execution runs indefinitely.

Finite state reachability graphs are the most common way of modeling systems. Reachability graphs, or Kripke structures, are designed as a graph where the nodes represent each reachable state within a system, while the edges represent the transitions between states [47]. To formally express the behavior of a system over time with respect to certain properties, one can use temporal logic. Linear temporal logic (LTL) is a common type of temporal logic where single execution paths can be checked [68]. Computation tree logic (CTL), on the other hand, expresses temporal logic in a tree structure, and thus, can reason about more than one time-line [39]. There are many other variations of temporal logic that could be used depending on the verification scenario [70]. Model checkers often leverage satisfiability modulo theories (SMT) solvers. They are used to solve satisfiability problems expressed in first-order logic.

Some other alternatives to theorem provers and model checkers include symbolic execution. In symbolic execution [33], symbols are used in lieu of the actual values of program variables or hardware components. As the simulation executes, the symbols can be updated, combined, turned into expressions, and sometimes simplified. Since the values of the symbols may not be clearly identified, each possible execution-branch is followed. In this way, all possible execution states can be evaluated simultaneously, at great cost of storage and slow execution during testing. There are variations of this technique, such as symbolic trajectory evaluation [90]. Unlike model checking, which requires a model of the system, symbolic execution deals directly with the source code, in case of software applications. The source code is executed in an interpreted environment that is able to keep track of all possible values of all variables, pointers, etc., in the code as it runs.

In summary, theorem provers or model checkers can be used for security verification of the secure processor architectures. Theorem provers have a much higher learning curve, meanwhile model checkers suffer from state explosion problem and in some cases may not be able to give a definite answer about security of a system.

9.4   DISCUSSION OF HARDWARE-SOFTWARE SECURITY VERIFICATION PROJECTS

Many of the existing security verification projects use general purpose verification tools. The current general-purpose tools used in security verification are not compatible with conventional hardware or software languages, such as C or Verilog, requiring either manual or programatic conversion of a design into a form that can be handled by the security verification tools.

Some project create a model that is separate from the actual system implementation source code, e.g., Micro-Policies [51], Cache verification [251], XOM [136], or SecVisor [66]. With these approaches, designers have to make sure their model accurately mirrors the system implementation, otherwise the result of verification might not be correct.

Some existing security verification projects take the approach of designing new domain-specific languages that allow making verification an integral part of the design and implementation process. In these projects, tools are developed to transform the system description in the new domain-specific language into another form that is amenable to use with verification tools, e.g., VeriCoq [25] or Dafny [131]. For example, in Dafny, the code has annotations for pre- and post-conditions, invariants, and ghost variables. With use of annotations and through automatic transformation, SMT solvers can check if the invariants always hold. Meanwhile, other projects embed security-related tags into a conventional language such as Verilog. These projects tend to develop custom tools to make sure the generated design has the desired security properties, these include SecChisel [54], Sapper [133], Caisson [134], or SecVerilog [250].

On the one hand, they can develop their system in a traditional language such as C or Verilog. This allows for quick development of the functional design with tools familiar to engineers, but does incur the effort of having to also separately write their design in a representation that their preferred security verifier understands. On the other hand, they can implement their system in a verification-friendly language. This has higher initial effort, but may pay off in long term with less effort due to not having to write the representation second time for verification. The drawback is that the verification-friendly language may not support all the aspects the designer desires to verify.

Confidentiality and integrity are the two main security properties often sought in a system (recall availability is not often provided at the level of single processor architecture). The verification aspect of a system often covers these properties, but can be formulated in a more generic form (e.g., non-interference) or a more specific form (e.g., memory integrity). The formulation of these properties depends on the levels of the system that are being verified, c.f. Figure 9.1, and on the tools used. The analysis of information flow provides is a useful basis for proving many of the security properties of a system. Monitoring information flow requires data labeling, declassification, and information flow rules specific to the system. Many projects focusing more on the hardware levels of a system use the analysis of information flow for proving information flow policies, non-interference, and confidentiality and integrity. Software-level focused projects have a wide variety of verification aspects, which try to verify confidentiality or integrity.

A selection of current and recent security verification projects focusing on hardware-levels of the system includes: Micro-Policies [51], Cache Verification [251], XOM [136], Veri-Coq [25], Formal-HDL [84], RTLIFT [184], Caisson [134], Sapper[133], SecVerilog [250], ReWire [171], and SecChisel [54].

Software-level focused security verification projects include: SeL4 [118], CetriKOS [80], Verve [246], Ironclad Apps [89], SecVisor [66], MinVisor [48], AAMP7G [235], ISA [190], MOAT [201], and Verification of SIR [200].

So far, formal verification research has been mostly focused on the functional correctness of the hardware or software systems. Security verification of software-only is also well studied. Hardware security verification, however, is an emerging research area which is necessitated by the fact that modern systems require both software and hardware for their correct and secure operation. Especially with the introduction of security-focused hardware, such as Intel SGX, trusting remote software and hardware is more critical now than before, as it handles users' ever-increasing sensitive information. Any vulnerabilities in these computing systems can be exploited by attackers. Thus, the whole system, including both the hardware and software parts, should be considered in the security verification.

Security verification is a branch of formal verification where the correctness properties are extended to include security properties, e.g., confidentiality and integrity. The process requires a formal specification of the security properties, an accurate representation of the implementation, and some verification mechanisms, e.g., theorem proving and model checking, to prove that the implementation complies with the needed security properties.

Since security properties are provided by system's components at multiple levels in the system, only verifying some particular level or levels cannot guarantee whole system's security. Especially, need to verify that microarchitecture correctly implements the architecture-specified security properties. With the improvement of verification tools and methods, more and more system levels should be included in the security verification.

In summary, there are many open research topics in the security verification of hardware and software systems. The most critical, however, is the need for full-system security verification, which spans more levels than can be done through today's existing projects.

9.5   SECURITY VERIFICATION ASSUMPTION

There are typically two assumptions relating to security verification of secure processors. At the end of the book, all the assumptions from all the book chapters are evaluated together.

9.5.1   VERIFIED TCB ASSUMPTION

The hardware and software TCB of a processor architecture should be verified both from functional and security perspective. Verification of the different modules and any protocols used for communication should be performed to ensure the TCB does not contain any logical flaws. Even with verification performed, there may still be issues later on at the manufacturing stage due to hardware trojans or supply chain attacks, for example.

9.5.2   VERIFIED TEE SOFTWARE ASSUMPTION

The software that is running inside the TEE should be verified. The TCB creates the trusted execution environment wherein the protected software executes. However, if the software is buggy or can be attacked, then there is noting the TCB can do to protect it.

CHAPTER 10

Principles of Secure Processor Architecture Design

This chapter concludes the book by presenting the principles of secure processor architecture design which have been derived based on the observations of best practices for secure processor design. It then gives a short description of how the secure design principles affect the standard computer architecture principles. It next overviews threats to the assumptions that have been discussed at the end of each chapter. It also presents common pitfalls and fallacies of secure processor architecture design. It ends with a list of challenges that secure processor architects can encounter, a list of future trends, and a brief note on the art and science of secure processor architecture design.

10.1 THE PRINCIPLES

For secure processor architecture, this book presents a set of five principles that can be followed to achieve a secure design:

1. protect off-chip communication and memory;

2. isolate processor state between TEE execution;

3. measure and continuously monitor TCB and TEE;

4. allow TCB introspection; and

5. minimize the TCB.

10.1.1 PROTECT OFF-CHIP COMMUNICATION AND MEMORY

Due to danger of attacks on external components, the secure processor chip assumption should be followed and any and all data and communication leaving the processor chip needs to be protected. It should be protected for confidentiality, integrity, access pattern, authentication and freshness. Encryption ensures protection of any secret or sensitive information (data or code) and should be used whenever data leaves the chip. Encryption should be used by the hardware and software parts of the TCB. Any memory location that is off-chip should be encrypted and hashed (using hash trees). Secure hash can be used to compute a fingerprint of piece of code and data stored outside the chip; the fingerprint can be used to ensure integrity. Randomized or oblivious accesses prevent attacks that use access patterns to learn information. All the off-chip communication and data needs to be authenticated with MACs or digital signatures, and freshness needs to be ensured (with proper use of nonces) so that attackers cannot use replay attacks against the system.

10.1.2 ISOLATE PROCESSOR STATE BETWEEN TEE EXECUTION

Within the processor, the hardware and software TCB create the TEE wherein protected software executes. Each TEE software instance shares the processor hardware with other instances of TEE software, and the untrusted software as well. Due to the shared hardware, numerous side channels or information leaks are possible. Any processor state should be isolated between execution of different TEE and also the untrusted software. The state includes caches, but also any buffers and memory structures in the processor, even if they are not visible to the programmers (such as the state of the branch predictor, for example). The state can be isolated through state flushing, where each time execution switches, all the state is securely stored, and flushed from the processor's functional units. State in other components, such as I/O devices, needs to be flushed as well.

10.1.3 MEASURE AND CONTINUOUSLY MONITOR TCB AND TEE

Measurement allows to attest the system and prove it is working correctly. Hashes and active measurements form basis of attestation. However, a reference hash or measurement value is always needed to be able to make a judgement about whether the state or behavior of the system is correct or not. Such reference values need to be provided for all TCB and TEE components. Attestation also may require use of public key infrastructure (PKI), thus key management and structure of the PKI needs to be designed when the secure processor architecture is designed.

Static hashing of code and data needs to be augmented with active measurement of the operation of the system to detect any run-time changes. Measurement can capture behavior of the system and detect any deviations. This should be applied both to the TCB itself, and the protected code and data in the TEE. Too often, TCB components are not actively measured, leaving a void in information about the status of the TCB. Integrity of all of the hardware and software parts of the TCB, and the protected code and data, should be ensured. Integrity checking can be augmented with error correction, to prevent both random and malicious changes to the code or data. Hashing is especially needed for any control data structures to ensure that information has not been modified. While the operation of the TCB should not be secret, to avoid security through obscurity, it still needs to be ensured that it is not modified.

10.1.4 ALLOW TCB INTROSPECTION

Transparency, or avoiding so-called security through obscurity, is needed to allow users to know what the system does. Also, attackers are often very clever, and a designer should not depend on hidden functionality to protect the system. Moreover, obscuring the functionality (e.g., making only binary code available) makes it difficult for legitimate users to be able to evaluate a system, or potentially fix a system if it is broken. Short-term benefits such as easier deployment is often offset if bugs are later found, exposing many systems to attacks.

10.1.5 MINIMIZE THE TCB

Having minimal TCB ensures that it can be validated and made trustworthy. There is a correlation between the complexity of the system (hardware and software lines of code) and number of potential bugs. Thus, minimizing TCB is crucial to reduce bugs, and consequently potential vulnerabilities that can lead to attacks. Also, for verification of the system, the simpler the system, the more likely it can be formally verified.

Any secure processor architecture relies on the TCB (a set of trusted hardware and software components) to realize the protections. However, a trusted component need not be trustworthy. A bug, or malicious modification, to a component in the TCB can render the system useless. Thus, formal security verification, and not just functional verification is needed. To enable security verification, TCB needs to remain small.

10.2 IMPACT OF SECURE DESIGN PRINCIPLES ON THE PROCESSOR ARCHITECTURE PRINCIPLES

In computer architecture there are roughly six principles: caching, pipelining, prediction, parallelization, use of indirection, and specialization. Most computer architecture techniques and improvements can be categorized into one of the six. The principles for secure processor architecture design have implications on the processor architecture principles. As can be seen from the discussion below, it is the principle of isolation of state that most often interacts with the existing architecture principles.

Caching. Principle of caching focuses on the idea that frequently used data should be cached and made available faster than some infrequently used data. This, however, creates fast and slow memory accesses that lead to side-channel attacks. Moreover, caches are a shared resource and have a limited storage capacity. The principle of isolation of state implies that two different TEEs should not affect each other. This may require flushing the caches to clean up data stored in the cache that would affect timing of memory accesses, or use of partitioning or randomization. The principle of isolation of state introduces performance overheads, which are at odds with the goal of caching that is to speed up memory accesses. Also, protection of off-chip memory implies use of encryption, which can further slow down memory accesses.

Pipelining. Pipelining breaks execution into small chunks (pipeline stages) and allows for exploitation of parallelism. To sustain high throughput of the processor pipelines prediction is used (the next principle discussed below). Prediction and the principle of isolation of state are at odds and the latter can potential limit how much prediction pipelining can use.

Prediction. Prediction aims to improve performance by predicting future behavior of the software (e.g., branch predictor predicts whether a branch should be taken or not taken even before the actual branch condition is evaluated). The principle of isolation of state implies that prediction should be made separately for each piece of TEE software. Meanwhile, many prediction approaches have a "global" component where the prediction is based in part on behavior of all the software. Any prediction mechanisms would need to consider each piece of software independently, and the performance impact is not clear—but likely slightly negative.

Parallelization. Parallelization focuses on performing multiple tasks in parallel. It may actually help isolation if different hardware units are available and not shared as different software executes in parallel. It may, however, make reasoning about behavior of the system (and thus measurement) more difficult as multiple events are happening in parallel at run time.

Use of Indirection. Page tables are one example of use of indirection where actual physical memory addresses are hidden and applications use virtual addresses. Indirection often uses small cache-like structures to speed up the lookup of the indirection information (e.g., look up page table translations in the TLB). Similar to the discussion on the caching principle, cache-like structure are at odds with the principle of isolation of state and the principle of isolation of state may have negative impact on performance of looking up of indirection information.

Specialization. Specialization is a catch-all architecture principle for any custom hardware or modules, e.g., special AES instructions and inclusion of AES engine in processors can be seen as specialization. If specialization involves any hardware units which contain state, then the state needs to be properly cleaned up or isolated between TEE execution due to the principle of isolation of state. This may have negative impact on performance of the specialized units.

10.3 LIMITATIONS OF THE SECURE PROCESSOR ASSUMPTIONS

Throughout the book, a number of assumptions that are made during secure processor architecture design were listed. They are repeated here, and the limitations of each assumption are briefly explained.

- Trusted Processor Chip Assumption—Invasive attacks, hardware trojans, and supply chain attacks can in practice mean that even verified design of a processor chip is not fully secure after it is manufactured.

- Small TCB Assumption—Code bloat and expansion of features present in secure processor architectures means that the TCB size keeps increasing.

- Open TCB Assumption—Proprietary code running on embedded security processors is one example of how the operation of the TCB is not open and exemplifies security through obscurity that can lead to attacks.

- No Side-effects Assumption—State in functional units is often not cleaned up properly, or there is processor state that was not considered (and thus not cleaned up) which leads to one piece of software being able to impact another; and this leads to side-channel attacks.

- Bug-free Protected Software Assumption—In practice, software running in TEE may not be verified or may contain bugs due to imported code modules.

- Trustworthy TCB Execution Assumption—Few, if any, architectures actually provide means to monitor the TCB execution, thus there is little information available to validate that TCB is trustworthy.

- Unique of Root of Trust Key Assumption—Most designs leverage keys burned in by the manufacturer, with little information about how manufacturer prevent others or themselves from cloning same key in multiple processors.

- Protected Root of Trust Assumption—For designs that leverage keys burned in by the manufacturer, little information is available on how the manufacturers actually protect the keys.

- Fresh Measurement Assumption—TOC-TOU attacks and no continuous measurement mean that measurements of the TCB or TEE may not be fresh in practice.

- Encrypted, Hashed, and Oblivious Access Memory Assumption—Lack of encryption, hashing or ORAM due to performance constraints means that data going off-chip to memory may be open to attacks, especially due to attackers observing access patterns.

- Protected Inter-processor Communication Assumption—Again, lack of encryption, hashing, or ORAM due to performance constraints means that communication can be vulnerable to attacks on confidentiality, integrity or access pattern observation.

- Side-Channel Free TEE Assumption—Lack of side channel protections or improper clean up of processor state means that there are possible side channels among TEE software and also the untrusted software.

- Verified TCB Assumption—Complexity of the TCB, and also the fact that it is composed of hardware and software parts, means it is very difficult to formally verify and few architectures today contain any formal security verification.

- Verified TEE Software Assumption—Small software running in a TEE is more likely to be verified, but constant code bloat makes the assumption hard to keep as the TEE software gets bigger and bigger.

10.4 PITFALLS AND FALLACIES

When designing secure processor architectures, researchers and practitioners may encounter some common pitfalls and fallacies. While it is hard to make an exhaustive list, below are some situations researchers and practitioners should pay attention to.

Pitfall: Security through Obscurity. The well-known idea of security through obscurity can lead to potential attacks and undermine the security of a processor architecture. There are numerous examples of designers hoping they are more clever than attackers, which have led to publicized security failures, e.g., [44]. It may be tempting to try to design custom-made cryptographic algorithms or protocols, e.g., to save on power or to improve performance of the system. Practice shows, however, that such custom cryptographic solutions often fail, and designers should stick with the true and tested ciphers such as AES for symmetric key cryptography or RSA for public key cryptography. Note, however, that even the true and tested ciphers should be used with care, especially with increasing concerns about quantum computers, post-quantum cryptographic algorithms should be considered, once they are standardized. Avoiding security through obscurity is also related to the idea behind Kerckhoffs's principle: security of a cryptographic system should only depend on the secret key, and not any secret designs of the system itself. This idea also applies beyond cryptographic algorithms or protocols, to any part of a secure processor design. Any hidden or secret functionality, often obfuscated or hidden due to intellectual property protection reasons, can be a source of vulnerability. Relying on hidden code or hardware design for any aspect of a secure processor design is a dangerous approach that should be avoided.

Fallacy: Assuming Hardware is Immutable. One of the motivations behind moving security features to hardware is that hardware is immutable, and thus more difficult for an attacker to probe or modify. This is true, mostly. As more and more functionality is moved to hardware, many features can be realized by software. Already, some secure processor designs include embedded processors, within the main processor, which are responsible for running security related code. Thus, the "hardware" features are simply code running on a processor that is, ideally, properly isolated from the rest of the actual processor. Now if this code can be modified, then the immutable aspect of the "hardware" is lost, opening potential attack vectors onto the secure processor design. The motivation, however, for adding some software to do what is assumed to be hardware's job can still be valid: as the complexity of the operations increases, creating custom state machines and actual hardware modules can be time-consuming and complex.

Designers should not force themselves to add all security features in hardware. They should be clear about which features are indeed in the immutable hardware logic gates, and which are implemented as code or microcode running on some embedded processor. Another aspect that architects should also consider is that hardware is composed of different physical pieces can be potentially modified by changing one of the pieces. One example here may be DRAM memory, which is separate from the main processor: both processor and memory may be designed to be secure, but one can easily swap in and out different memory modules, so the whole system has to protect against some pieces being replaced or modified.

Pitfall: Wrong Threat Model. The choice of a threat model is inherently a subjective decision. There is no perfect threat model. While it is important to make arguments about threat models, researchers and practitioners should distinguish attacking threat model vs. attacking the design. Especially, it may be possible to find a flaw with a threat model, and demonstrate an attack against the system under a threat that it was not originally designed for. In this case, the new attack is valid because the attacker found or designed, hopefully, reasonable new threat model that should be considered. Meanwhile, the original design is still correct, as it was not meant to protect against the original threat model, so the new attack is null when considering the modified threat model. When a flaw is found with a design, the attackers should clarify if they are attacking the design, or really if they are disputing the original threat model. At the same time, designers should acknowledge shortcomings of their threat models, and update the architectures if a new, better threat model is found.

Pitfall: Outdated or Custom Crypto. Already mentioned as part of the first pitfall, but worth highlighting again, both custom cryptographic protocols and algorithms should not be used in a secure processor architecture. Architecture design is already a complex process, and adding another layer of complexity, the unverified cryptographic protocols or algorithms only makes the design less reliable. The designers should, however, ensure that when new cryptographic standard is developed and agreed upon, the design is updated to use it. For example, DES has been superseded by AES, so secure processor designs should not use DES anymore. If or when a standard cipher such as AES or RSA is superseded, the secure processor design should be updated accordingly. Moreover, as updating hardware is expensive, use of known-good cryptographic algorithms ensures that the processors are likely secure for many years to come and do not require sudden replacement due to a flaw in the algorithm.

Pitfall: Ignoring Side Channels. New side-channel attacks, as well as covert-channel attacks, are discovered each year. While side-channel attacks have much less bandwidth than other types of attacks, they can still be used to steal cryptographic keys, which maybe sufficient to undermine security of a system. Addressing side channels is costly, often requires a redesign and a performance degradation of the system. Nevertheless, with move to cloud computing (where many mutually-untrusting users may share same computer and be exposed to software-based attacks) or Internet-of-Things (where low-cost devices with few security features may be exposed to physical attacks), there is need to consider, and protect against side- and covert-channel attacks.

Pitfall: Incorrect Hardware Abstractions. Use of abstraction, such as having the ISA as the hardware-software interface, helps to design and reason about operation of the system. However, when the abstraction does not match how the real device or hardware behaves, then security attacks are possible. Many security attacks can be attributed to use of wrong abstractions, or abstractions not matching reality. Coldboot [86] or Rowhammer [117] attack, for example, exemplify how most computer architects assumed that data in DRAM is volatile and is lost immediately after power is turned off, or that the only way to modify some data in particular DRAM address is to explicitly write to to that address, respectively. Meanwhile, for both of the attacks, physical properties of the hardware differed from the abstract view of DRAM, and allowed these attacks to happen. In the processor chip itself, attacks such as Spectre [120] and Meltdown [138], for example, exemplify how at the ISA level there is assumption that there are no visible state changes if instructions executed through speculative execution are invalidated. Meanwhile, at the microarchitecture level, speculative execution was shown to modify the processor cache state, and that state was not cleaned up. Designers should ensure the abstraction they use matches the real hardware, and if needed new or extended abstractions should be designed. Especially, it is important to expose security-related behavior of the system in the abstraction.

Fallacy: Assuming a System if Fully Secure. It is unlikely that a practical implementation of a system will ever be fully secure. Even if the design is fully correct, then there are issues of implementation and physical realization of the system. Also, even if the designer is confident that there are no flaws, a clever attacker can often find a way around—attacker only needs to find one way into the system, while the defender has to protect from all the possible attacks he or she can think of. In addition, if formal verification is used, it is likely impossible to prove absence of any attack. Bearing this in mind, architects should assume that a system will be attacked. Some consequences are that devices should not share same, hard-coded cryptographic keys, lest one of them is compromised, or that continuous attestation should be used and it may help detect attacks, rather than doing attestation only once at system startup.

Pitfall: Ignoring Security as Primary Design Objective. Often, performance, area, and energy are the primary design targets for a processor architecture. This can lead to design choices that may sometimes undermine security. Most prominent examples of where security issues can arise from prioritizing other design choices include the variety of timing side-channel attacks. For example, the processor hardware is shared among different programs (to save area), but this leads to contention-based attacks, or processor state is lazily cleaned up (to improve performance), but this leads to data leakage through registers that are not cleaned up. Performance, area, and energy should be considered, but not independent of security. Zero-overhead security is something that should still be strived for, and perhaps one day will be achieved.

Pitfall: Code Bloat. Most of the secure processor architectures are designed such that they protect some code and data. This to-be protected code and data is assumed to be itself secure. To help ensure that the protected code and data is secure, one approach is to minimize it, just as one wants to minimize the TCB. However, over time, there is code bloat. Once system designers discover the benefits of some new security feature, e.g., Intel's SGX, they are enticed to put more and more code to execute inside the protected environment. Eventually, while SGX may work, it is not possible to ensure that the code inside the SGX enclaves is correct. Code bloat can also affect the secure processor's hardware and software. As users demand more and more options, the complexity of the TCB grows. This in turn increases the size of the hardware and software code, leading to bugs, and many have been discovered in commercial products.

10.5   CHALLENGES IN SECURE PROCESSOR DESIGN

The research area of computer architecture and security is quickly evolving. Nevertheless, there are a number of challenges, listed below in alphabetical order, that are recurring over the years. Understanding these challenges and addressing them could help in design of better secure processor architectures of tomorrow.

Attack Recovery. Recovering from, or responding to, an attack is an open challenge. The easiest solution in secure processor architectures has been to halt or shut down the system. However, such action may not be feasible in real-life deployments, e.g., if a secure processor is part of a self-driving car it cannot simply stop working when there is an attack. Degrading system performance is one option. However, in general, the practical issue of what to do after an attack is detected is an open research challenge.

Concurrency. Use of many processor cores, or even speculation mechanisms in a single core, means that reasoning about timing of different events is difficult. Especially, security checks need to complete before data is accessed; or if speculative actions were taken, all side-effects of such actions need to be removed in case of failed security checks.

Data Remanence. Data retention, or unexpected existence there of, can cause problems for secure processor designs. One of the well-known attacks is the Coldboot attack where researchers proved that DRAM memories can retain data for a long time, even after removed from the computer. This showed dangers that potentially sensitive data, through to disappear once computer is powered off, was able to be accessed. More recently, non-volatile memory, NVRAM, is a growing trend in computer architectures, which also leads to data retention related issues: data does not disappear when system is turned off. The intended, or unintended, persistence of data means that encryption needs to be used to secure the data in case it is not deleted at some expected time.

Key Management. Management of cryptographic keys is always a difficult challenge. Often, academic publications omit it, as it requires considering many more components than just the secure system, e.g., manufacturer, other systems, etc. Nevertheless, this is often the weak point, or at least potentially something that prevents the system from being deployed in real life.

Threat Models and Economics. The economic part of security is often a hidden topic in area of architecture security. Meanwhile, security is really about economics. Why assume that physical attack is not feasible while software attacks are okay? Such questions about the threat model often can boil down to some implied assumptions about how many resources and money the attacker would, or would not, use to break a system. If it is valuable enough, the attacker's resources could be huge and most defenses will fail. Understanding and motivating architecture security work in economic terms could perhaps help better define existing threat models. Moreover, risk assessment and risk management can help teach architects about how to designs systems and what threats to expect.

Time of Check to Time of Use. The so-called TOCTOU attacks are a large class of problems relating to attestation. The difficulty of these problems is that at the time of check when some attestation information is gathered a system is operating correctly and not under attack; meanwhile, by the time the attestation information is actually used, the time of use, the system has been compromised—this is not easily discovered as the attestation information is from the time of check, which has already passed. In general, attestation information is gathered at system startup, when it is easy to reason about what is the correct initial state. However, as the system runs, it changes it state and defining a correct state at any point in time is difficult. Active, run-time protections and attestation are one possible solution.

Usability. If a feature is not user-friendly, or difficult to use, people will find a way around it, or not use it at all. One such challenge is with attestation: it is easy to generate a hash of the code, however, conveying to users why certain hash value, a random looking string of digits, is good and another, also random looking string, is not, is difficult. Even if users agree on the known-good values, any small change to software changes the hash, so practically updating the system is not possible as the number of hashes to be managed grows quickly over time. Considering users, and how easy the system is to use in practice is important. A related issue is of compatibility, and whether code has to be modified or re-compiled. Many excellent research ideas may not get deployed quickly, or at all, if they necessitate too many changes to deployed systems. There is a tradeoff between how many security features one can add, e.g., using a clean-slate approach, and the effort to get such a system to be deployed in practice.

10.6   FUTURE TRENDS IN SECURE PROCESSOR DESIGNS

While predicting the future is fraught with danger, there are some near- and long-term research directions that may be interesting for computer architects to consider.

Homomorphic Encryption. Practical homomorphic encryption is always few years away, but steadily it is getting nearer practice. It could be a big change in design of secure processor architectures. Especially, there is no longer need to trust the manufacturer. Performance, however, remains the main hurdle in actually implementing it in commodity processors.

System Designs Leveraging TEEs. With the recent deployment of trusted execution environments such as Intel's SGX, AMD's SEV, or ARM's TrustZone, there are actual secure processors designs people can use. There is a shift now toward more development of system software and applications that leverage these hardware features. Already, conferences or workshops dedicated to SGX-related software exist. Meanwhile, architects should observe the needs of the software and large systems, that use the secure hardware as one of many building blocks, to explore what new features can be used to help the system design. Focusing on just the secure processor architectures may no-longer be sufficient, rather the focus may shift to larger system design, were the processor is just one of the components.

Heterogenous Systems. Traditionally, secure processor designs have focused on just the processor (and the memory subsystem that often uses external DRAM chips). Today's computers, however, are more and more heterogenous. Most systems include GPU and now FPGA chips. Protections offered by the secure processors will need to be extended to consider GPUs and FPGAs, and how to provide confidentiality and integrity protection to software and hardware designs that span code running on CPU, GPU, with custom logic that is being constantly reconfigured in the FPGA fabric.

Quantum Computers. Always few years away, but always getting closer to reality, quantum computers offer a promise and challenge for secure processor architectures. On one hand, new processing will be designed around quantum computers. Early work is today already exploring how to design processor architectures for quantum computers. Such quantum computer-based architectures will naturally evolve to include security. Designers should learn from classical architectures (e.g., about issues such as side-channel attacks) and aim to prevent them already in early days of designing quantum computer based architectures. On the other hand, quantum computers are known to be able to break certain cryptographic algorithms. Thus, secure processor designers (and especially industry) need to ensure that cryptographic algorithms used are sufficiently strong; computer manufactured today will be in use for many years, and practical quantum computers may come online in that time-frame.

Physically Uncloneable Functions. Leveraging unique physical hardware features due to manufacturing differences has emerged as important research area in hardware security. The so-called PUF, extract a hardware fingerprint (not to be confused with hash functions and the hash values which are also akin to fingerprints), can be used for secure key storage, or even for random number generation. Architects can leverage the PUFs in the secure processor designs for these functionalities. The measurement of the physical properties could be, for example, used to attest the operation of the hardware, and combine it with the software attestation operations. Already, researchers are exploring aging and means to test how long certain hardware has been on and running.

Non-CMOS-based Architectures. Secure processor architectures have been traditionally designed based on CMOS technologies. However, new types of devices (memristor, ferroelectric-based, DNA-based computing, etc.) are being now explored. The secure processor designers can leverage these technologies to design new and better designs. The properties of these devices need to be studied to, especially focusing on side channels, to ensure they will no have some unintended side-effects that can leak information.

10.7   ART AND SCIENCE OF SECURE PROCESSOR DESIGN

Raising the bar for the attackers is what most of the security is about. Anything can be attacked and 100% security does not exist. But, if the difficulty is high enough, most attacks will fail. Secure processor architecture design is about developing systems that can protect code and data executing on a commodity-style processor while protecting from most practical attacks. The art of secure processor architecture is about how to pick the threat models and what protections to offer, while the science is about what algorithms and hardware (and software) modules to use to implement these protections efficiently without sacrificing too much performance, power, or area of the processor. It is hoped that through learning about the secure processor architectures and the principles for designing them, the readers have gained some insights into the art and science of secure processor design.

Bibliography

[1] Onur Acıiçmez and Çetin Kaya Koç. Trace-driven Cache Attacks on AES (short paper). In *Information and Communications Security*, pages 112–121. Springer, 2006. DOI: 10.1007/11935308_9 10, 91

[2] Onur Aciiçmez, Çetin Kaya Koç, and Jean-Pierre Seifert. On the Power of Simple Branch Prediction Analysis. In *Proceedings of the Symposium on Information, Computer and Communications Security*, pages 312–320. ACM, 2007. 101

[3] Onur Acıiçmez, Werner Schindler, and Çetin K Koç. Cache Based Remote Timing Attack on the AES. In *Topics in Cryptology*, pages 271–286. Springer, 2006. DOI: 10.1007/11967668_18 91

[4] Onur Acıiçmez, Werner Schindler, and Çetin K Koç. Cache Based Remote Timing Attack on the AES. In *Cryptographers' Track at the RSA Conference*, pages 271–286. Springer, 2007. DOI: 10.1007/11967668_18 10

[5] Keith Adams and Ole Agesen. A Comparison of Software and Hardware Techniques for x86 Virtualization. *ACM SIGOPS Operating Systems Review*, 40(5):2–13, 2006. DOI: 10.1145/1168918.1168860 31, 70

[6] Dakshi Agrawal, Bruce Archambeault, Josyula R. Rao, and Pankaj Rohatgi. The EM Side-Channels. In *International Workshop on Cryptographic Hardware and Embedded Systems*, pages 29–45. Springer, 2002. DOI: 10.1007/3-540-36400-5_4 10, 41

[7] Jeongseob Ahn, Seongwook Jin, and Jaehyuk Huh. Revisiting Hardware-Assisted Page Walks for Virtualized Systems. In *ACM SIGARCH Computer Architecture News*, volume 40, pages 476–487. IEEE Computer Society, 2012. DOI: 10.1145/2366231.2337214 31

[8] Kahraman Akdemir, Martin Dixon, Wajdi Feghali, Patrick Fay, Vinodh Gopal, Jim Guilford, Erdinc Ozturk, Gil Wolrich, and Ronen Zohar. Breakthrough AES Performance with Intel AES New Instructions, White Paper, June 2010. https://software.intel.com/sites/default/files/m/d/4/1/d/8/10TB24_Breakthrough_AES_Performance_with_Intel_AES_New_Instructions.final.secure.pdf. 80, 82

[9] NEC Akkaya, B. Erbagci, and K. Mai. Combatting IC Counterfeiting Using Secure Chip Odometers. In *Proceedings of the IEEE International Electron Devices Meeting*, pages 39–5. IEEE, 2017. DOI: 10.1109/iedm.2017.8268523 41, 60

[10] AMD Secure Processor. www.amd.com/en-us/innovations/software-technologies /security. 33

[11] AMD. AMD Memory Encryption, 2016. http://amd-dev.wpengine.netdna-cdn.com/wordpress/media/2013/12/AMD_Memory_Encryption_Whitepaper_v7-Public.pdf. 103

[12] Secure Encrypted Virtualization Key Management, 2017. https://support.amd.com/TechDocs/55766_SEV-KM%20API_Specification.pdf. 35, 36, 70

[13] Ittai Anati, Shay Gueron, Simon Johnson, and Vincent Scarlata. Innovative Technology for CPU Based Attestation and Sealing. In *Proceedings of the Workshop on Hardware and Architectural Support for Security and Privacy*, 2013. 34, 35, 36, 39, 48, 72

[14] Ross J. Anderson. *Security Engineering: a Guide to Building Dependable Distributed Systems*. John Wiley & Sons, 2010. 21, 41

[15] Sergei Arnautov, Bohdan Trach, Franz Gregor, Thomas Knauth, Andre Martin, Christian Priebe, Joshua Lind, Divya Muthukumaran, Dan O'Keeffe, Mark Stillwell, et al. SCONE: Secure Linux Containers with Intel SGX. In *Proceedings of the USENIX Symposium on Operating Systems Design and Implementation*, pages 689–703, 2016. 51

[16] Amittai Aviram, Sen Hu, Bryan Ford, and Ramakrishna Gummadi. Determining Timing Channels in Compute Clouds. In *Proceedings of the Workshop on Cloud Computing Security Workshop*, pages 103–108. ACM, 2010. DOI: 10.1145/1866835.1866854 101

[17] Ahmed M. Azab, Peng Ning, and Xiaolan Zhang. Sice: a Hardware-Level Strongly Isolated Computing Environment for x86 Multi-Core Platforms. In *Proceedings of the Conference on Computer and Communications Security*, pages 375–388. ACM, 2011. DOI: 10.1145/2046707.2046752 51

[18] Jonathan Bachrach, Huy Vo, Brian Richards, Yunsup Lee, Andrew Waterman, Rimas Avižienis, John Wawrzynek, and Krste Asanović. Chisel: Constructing Hardware in a Scala Embedded Language. In *Proceedings of the Annual Design Automation Conference*, pages 1216–1225. ACM, 2012. DOI: 10.1145/2228360.2228584 107

[19] Davide B. Bartolini, Philipp Miedl, and Lothar Thiele. On the Capacity of Thermal Covert Channels in Multicores. In *Proceedings of the European Conference on Computer Systems*, page 24. ACM, 2016. DOI: 10.1145/2901318.2901322 29, 96

[20] Béatrice Bérard, Michel Bidoit, Alain Finkel, François Laroussinie, Antoine Petit, Laure Petrucci, and Philippe Schnoebelen. *Systems and Software Verification: Model-Checking Techniques and Tools.* Springer Science & Business Media, 2013. DOI: 10.1007/978-3-662-04558-9. 104

[21] Daniel J. Bernstein. Cache-timing attacks on AES, 2005. `https://cr.yp.to/antifor gery/cachetiming-20050414.pdf` 10, 96

[22] Daniel J. Bernstein. ChaCha, a Variant of Salsa20. In *Workshop Record of SASC*, volume 8, pages 3–5, 2008. 17

[23] Daniel J. Bernstein, Johannes Buchmann, and Erik Dahmen, editors. *Post-Quantum Cryptography.* Springer, Heidelberg, 2009. DOI: 10.1007/978-3-540-88702-7 18

[24] Ravi Bhargava, Benjamin Serebrin, Francesco Spadini, and Srilatha Manne. Accelerating Two-Dimensional Page Walks for Virtualized Systems. In *ACM SIGARCH Computer Architecture News*, volume 36, pages 26–35. ACM, 2008. DOI: 10.1145/1353534.1346286 31

[25] Mohammad-Mahdi Bidmeshki and Yiorgos Makris. VeriCoq: A Verilog-to-Coq Converter for Proof-Carrying Hardware Automation. In *Proceedings of the International Symposium on Circuits and Systems*, pages 29–32. IEEE, 2015. DOI: 10.1109/iscas.2015.7168562 104, 110

[26] Andrey Bogdanov, Thomas Eisenbarth, Christof Paar, and Malte Wienecke. Differential Cache-Collision Timing Attacks on AES with Applications to Embedded CPUs. In *Topics in Cryptology*, volume 10, pages 235–251. Springer, 2010. DOI: 10.1007/978-3-642-11925-5_17 91

[27] Andrey Bogdanov, Lars R. Knudsen, Gregor Leander, Christof Paar, Axel Poschmann, Matthew J. B. Robshaw, Yannick Seurin, and Charlotte Vikkelsoe. PRESENT: An Ultra-Lightweight Block Cipher. In *International Workshop on Cryptographic Hardware and Embedded Systems*, pages 450–466. Springer, 2007. DOI: 10.1007/978-3-540-74735-2_31 17

[28] Joseph Bonneau and Ilya Mironov. Cache-Collision Timing Attacks Against AES. In *Cryptographic Hardware and Embedded Systems*, pages 201–215. Springer, 2006. DOI: 10.1007/11894063_16 91, 96

[29] Ernie Brickell, Jan Camenisch, and Liqun Chen. Direct Anonymous Attestation. In *Proceedings of the Conference on Computer and Communications Security*, pages 132–145. ACM, 2004. DOI: 10.1145/1030083.1030103 23, 63

[30] Billy Bob Brumley and Risto M. Hakala. Cache-Timing Template Attacks. In *Advances in Cryptology–ASIACRYPT 2009*, pages 667–684. Springer, 2009. DOI: 10.1007/978-3-642-10366-7_39 91

[31] J. Burns and J.-L. Gaudiot. SMT Layout Overhead and Scalability. *IEEE Transactions of Parallel and Distributed Systems*, 13(2):142–155, Feb 2002. DOI: 10.1109/71.983942 90

[32] Jan Camenisch, Simone Fischer-Hübner, and Kai Rannenberg. *Privacy and Identity Management for Life*. Springer Science & Business Media, 2011. DOI: 10.1007/978-3-642-20317-6 21

[33] William C. Carter, William H. Joyner Jr., and Daniel Brand. Symbolic Simulation for Correct Machine Design. In *Conference on Design Automation*, pages 280–286. IEEE, 1979. DOI: 10.1109/dac.1979.1600119 109

[34] Championship Branch Prediction, 2014. http://www.jilp.org/cbp2014/. 91

[35] David Champagne and Ruby B. Lee. Scalable Architectural Support for Trusted Software. In *Proceedings of the International Symposium on High Performance Computer Architecture*, 2010. DOI: 10.1109/hpca.2010.5416657 34, 35, 36, 38, 47, 48, 72

[36] David Champagne and Ruby B. Lee. Scalable Architectural Support for Trusted Software. In *International Symposium on High Performance Computer Architecture*, pages 1–12. IEEE, 2010. DOI: 10.1109/hpca.2010.5416657 103, 105

[37] Lily Chen, Dustin Moody, and Yi-Kai Liu. NIST post-quantum cryptography standardization, 2017. https://csrc.nist.gov/projects/post-quantum-cryptography/post-quantum-cryptography-standardization/. 18

[38] Siddhartha Chhabra, Brian Rogers, Yan Solihin, and Milos Prvulovic. Making Secure Processors OS-and Performance-Friendly. *ACM Transactions on Architecture and Code Optimization*, 5(4), 2009. DOI: 10.1145/1498690.1498691 72

[39] Edmund M. Clarke and E. Allen Emerson. Design and Synthesis of Synchronization Skeletons using Branching Time Temporal Logic. In *Proceedings of the Workshop on Logic of Programs*, pages 52–71. Springer, 1981. DOI: 10.1007/bfb0025774 109

[40] Pat Conway and Bill Hughes. The AMD Opteron Northbridge Architecture. *IEEE Micro*, 27(2), 2007. DOI: 10.1109/mm.2007.43 80

[41] Stephen A. Cook. The Complexity of Theorem-Proving Procedures. In *Proceedings of the ACM Symposium on Theory of Computing*, pages 151–158. ACM, 1971. DOI: 10.1145/800157.805047 108

[42] CoqIde. https://github.com/coq/coq/wiki/CoqIde. 108

[43] Victor Costan, Ilia A. Lebedev, and Srinivas Devadas. Sanctum: Minimal Hardware Extensions for Strong Software Isolation. In *Proceedings of the USENIX Security Symposium*, pages 857–874. USENIX Association, 2016. 38, 48, 100

[44] Nicolas Courtois, Karsten Nohl, and Sean O'Neil. Algebraic Attacks on the Crypto-1 Stream Cipher in MiFare Classic and Oyster Cards. Cryptology ePrint Archive, Report 2008/166, 2008. https://eprint.iacr.org/2008/166. 7, 17, 118

[45] CPU Bugs. https://wiki.osdev.org/CPU_Bugs. 29

[46] Karl Crary. *Toward a Foundational Typed Assembly Language.* ACM, 2003. DOI: 10.1145/604131.604149 107

[47] Max J. Cresswell and George Edward Hughes. *A New Introduction to Modal Logic.* Routledge, 2012. DOI: 10.4324/9780203290644 109

[48] Mike Dahlin, Ryan Johnson, Robert Bellarmine Krug, Michael McCoyd, and William Young. Toward the Verification of a Simple Hypervisor. *arXiv preprint arXiv:1110.4672*, 2011. DOI: 10.4204/eptcs.70.3 110

[49] DARPA. Leveraging the Analog Domain for Security (LADS). https://www.darpa.mil/program/leveraging-the-analog-domain-for-security. 102

[50] Tom Woller, David Kaplan, and Jeremy Powell. AMD Memory Encryption, White Paper, April 2016. http://developer.amd.com/wordpress/media/2013/12/AMD_Memory_Encryption_Whitepaper_v7-Public.pdf. 34, 39, 49, 70

[51] Arthur Azevedo De Amorim, Maxime Dénes, Nick Giannarakis, Catalin Hritcu, Benjamin C. Pierce, Antal Spector-Zabusky, and Andrew Tolmach. Micro-Policies: Formally Verified, Tag-Based Security Monitors. In *Proceedings of the Symposium on Security and Privacy*, pages 813–830. IEEE, 2015. DOI: 10.1109/sp.2015.55 104, 109, 110

[52] Onur Demir, Wenjie Xiong, Faisal Zaghloul, and Jakub Szefer. Survey of Approaches for Security Verification of Hardware/Software Systems. Cryptology ePrint Archive, Report 2016/846, 2016. http://eprint.iacr.org/2016/846. 6, 11, 12, 103, 104

[53] Shuwen Deng, Doğuhan Gümüşoğlu, Wenjie Xiong, Y. Serhan Gener, Onur Demir, and Jakub Szefer. SecChisel: Language and Tool for Practical and Scalable Security Verification of Security-Aware Hardware Architectures. Cryptology ePrint Archive, Report 2017/193, 2017. https://eprint.iacr.org/2017/193. 104

[54] Shuwen Deng, Doğuhan Gümüşoğlu, Wenjie Xiong, Y. Serhan Gener, Onur Demir, and Jakub Szefer. SecChisel: Language and Tool for Practical and Scalable Security Verification of Security-Aware Hardware Architectures. Cryptology ePrint Archive, Report 2017/193, 2017. https://eprint.iacr.org/2017/193. 107, 110

[55] Srinivas Devadas, Edward Suh, Sid Paral, Richard Sowell, Tom Ziola, and Vivek Khandelwal. Design and Implementation of PUF-Based "Unclonable" RFID ICs for Anti-Counterfeiting and Security Applications. In *Proceedings of the International Conference on RFID*, pages 58–64. IEEE, 2008. DOI: 10.1109/rfid.2008.4519377 40

[56] Leonid Domnitser, Aamer Jaleel, Jason Loew, Nael Abu-Ghazaleh, and Dmitry Pono-marev. Non-Monopolizable Caches: Low-Complexity Mitigation of Cache Side Chan-nel Attacks. *ACM Transactions on Architecture and Code Optimization*, 8(4):35, 2012. DOI: 10.1145/2086696.2086714 100

[57] Morris J. Dworkin. SHA-3 Standard: Permutation-Based Hash and Extendable-Output Functions. Technical report, NIST, 2015. DOI: 10.6028/nist.fips.202 21

[58] Jeffrey S. Dwoskin and Ruby B. Lee. Hardware-Rooted Trust for Secure Key Management and Transient Trust. In *Proceedings of the Conference on Computer and Communications Security*, pages 389–400. ACM, 2007. DOI: 10.1145/1315245.1315294 36, 38, 47, 48, 103

[59] Shawn Embleton, Sherri Sparks, and Cliff C. Zou. SMM Rootkit: A New Breed of OS Independent Malware. *Security and Communication Networks*, 6(12):1590–1605, 2013. DOI: 10.1002/sec.166 33, 50, 51

[60] Mark Ermolov and Maxim Goryachy. How to Hack a Turned-Off Computer, or Running Unsigned Code in Intel Management Engine. *Black Hat Europe*, 2017. 7, 34, 39

[61] Dmitry Evtyushkin, Dmitry Ponomarev, and Nael Abu-Ghazaleh. Covert Channels Through Branch Predictors: a Feasibility Study. In *Proceedings of the Fourth Workshop on Hardware and Architectural Support for Security and Privacy*, page 5. ACM, 2015. DOI: 10.1145/2768566.2768571 96

[62] Dmitry Evtyushkin, Dmitry Ponomarev, and Nael Abu-Ghazaleh. Understanding and Mitigating Covert Channels Through Branch Predictors. *ACM Transactions on Architecture and Code Optimization (TACO)*, 13(1):10, 2016. DOI: 10.1145/2870636 96

[63] Christopher W. Fletcher, Ling Ren, Albert Kwon, Marten Van Dijk, Emil Stefanov, Dim-itrios Serpanos, and Srinivas Devadas. A Low-Latency, Low-Area Hardware Oblivious Ram Controller. In *IEEE 23rd Annual International Symposium on Field-Programmable Custom Computing Machines*, pages 215–222. IEEE, 2015. DOI: 10.1109/fccm.2015.58 74

[64] Christopher Wardlaw Fletcher. *Ascend: An Architecture for Performing Secure Computation on Encrypted Data*. PhD thesis, Massachusetts Institute of Technology, Department of Electrical Engineering and Computer Science, 2013. 36, 38, 48

[65] Caroline Fontaine and Fabien Galand. A Survey of Homomorphic Encryption for Nonspecialists. *EURASIP Journal of Information Security*, 2007, January 2007. DOI: 10.1155/2007/13801 36

[66] Jason Franklin, Arvind Seshadri, Ning Qu, Sagar Chaki, and Anupam Datta. Attacking, Repairing, and Verifying SecVisor: A Retrospective on the Security of a Hypervisor. Technical report, Technical Report CMU-CyLab-08-008, Carnegie Mellon University, 2008. 109, 110

[67] Felix C. Freiling and Sebastian Schinzel. Detecting Hidden Storage Side Channel Vulnerabilities in Networked Applications. In *IFIP International Information Security Conference*, pages 41–55. Springer, 2011. DOI: 10.1007/978-3-642-21424-0_4 9

[68] Dov Gabbay, Amir Pnueli, Saharon Shelah, and Jonathan Stavi. On the temporal analysis of fairness. In *Proceedings of the Symposium on Principles of Programming Languages*, pages 163–173. ACM, 1980. DOI: 10.1145/567446.567462 109

[69] John E. Gaffney. Estimating the Number of Faults in Code. *IEEE Transactions on Software Engineering*, 4:459–464, 1984. DOI: 10.1109/tse.1984.5010260 2, 5

[70] Iván Garcıa-Ferreira, Carlos Laorden, Igor Santos, and Pablo Garcia Bringas. A Survey on Static Analysis and Model Checking. In *International Joint Conference SOCO'14-CISIS'14-ICEUTE'14*, page 443, 2014. DOI: 10.1007/978-3-319-07995-0_44 109

[71] Philip Garrou, Christopher Bower, and Peter Ramm. *Handbook of 3D Integration: Volume 1 – Technology and Applications of 3D Integrated Circuits*. John Wiley and Sons, 2011. DOI: 10.1002/9783527623051. 81

[72] Blaise Gassend, G Edward Suh, Dwaine Clarke, Marten Van Dijk, and Srinivas Devadas. Caches and Hash Trees for Efficient Memory Integrity Verification. In *Proceedings of the International Symposium on High-Performance Computer Architecture*, pages 295–306. IEEE, 2003. DOI: 10.1109/hpca.2003.1183547 71

[73] Craig Gentry. Fully Homomorphic Encryption Using Ideal Lattices. In *Proceedings of the Symposium on Theory of Computing*, pages 169–178, 2009. DOI: 10.1145/1536414.1536440 42

[74] Oded Goldreich. Towards a Theory of Software Protection and Simulation by Oblivious RAMs. In *Proceedings of the Symposium on Theory of Computing*, pages 182–194. ACM, 1987. DOI: 10.1145/28395.28416 73

[75] Johannes Götzfried, Moritz Eckert, Sebastian Schinzel, and Tilo Müller. Cache Attacks on Intel SGX. In *Proceedings of the European Workshop on Systems Security*. ACM, 2017. DOI: 10.1145/3065913.3065915 12

[76] Philipp Grabher, Johann Großschädl, and Daniel Page. Cryptographic Side-Channels from Low-Power Cache Memory. In *Cryptography and Coding*, pages 170–184. Springer, 2007. DOI: 10.1007/978-3-540-77272-9_11 91

[77] James W. Gray III. On Introducing Noise Into the Bus-Contention Channel. In *Proceedings of the IEEE Computer Symposium Symposium on Research in Security and Privacy*, pages 90–98. IEEE, 1993. DOI: 10.1109/risp.1993.287640 101

[78] Lov K. Grover. A Fast Quantum Mechanical Algorithm for Database Search. In *Symposium on the Theory of Computing*, pages 212–219. ACM, 1996. DOI: 10.1145/237814.237866 18

[79] Adam Grummitt. *Capacity Management-A Practitioner Guide*. Van Haren, 1970. 31, 39

[80] Ronghui Gu, Jérémie Koenig, Tahina Ramananandro, Zhong Shao, Xiongnan Newman Wu, Shu-Chun Weng, Haozhong Zhang, and Yu Guo. Deep specifications and certified abstraction layers. In *ACM SIGPLAN Notices*, volume 50, pages 595–608. ACM, 2015. DOI: 10.1145/2775051.2676975 104, 110

[81] Jorge Guajardo, Sandeep S. Kumar, Geert-Jan Schrijen, and Pim Tuyls. *FPGA Intrinsic PUFs and Their use for IP Protection*. Springer, 2007. DOI: 10.1007/978-3-540-74735-2_5 23, 61

[82] Jorge Guajardo, Sandeep S Kumar, Geert Jan Schrijen, and Pim Tuyls. Brand and IP Protection with Physical Unclonable Functions. In *Proceedings of the International Symposium on Circuits and Systems*, pages 3186–3189. IEEE, 2008. DOI: 10.1109/iscas.2008.4542135 23, 61

[83] David Gullasch, Endre Bangerter, and Stephan Krenn. Cache Games – Bringing Access-Based Cache Attacks on AES to Practice. In *Proceedings of the Symposium on Security and Privacy*, pages 490–505. IEEE, 2011. DOI: 10.1109/sp.2011.22 10

[84] Xiaolong Guo, Raj Gautam Dutta, Prabhat Mishra, and Yier Jin. Automatic RTL-to-Formal Code Converter for IP Security Formal Verification. In *Proceedings of the International Workshop on Microprocessor and SOC Test and Verification*, pages 35–38. IEEE, 2016. DOI: 10.1109/mtv.2016.23 110

[85] Aarti Gupta. Formal Hardware Verification Methods: a Survey. In *Computer-Aided Verification*, pages 5–92. Springer, 1992. DOI: 10.1007/978-1-4615-3556-0_2 104

[86] J. Alex Halderman, Seth D. Schoen, Nadia Heninger, William Clarkson, William Paul, Joseph A. Calandrino, Ariel J. Feldman, Jacob Appelbaum, and Edward W. Felten. Lest We Remember: Cold-Boot Attacks on Encryption Keys. *Communications of the ACM*, 52(5):91–98, 2009. DOI: 10.1145/1506409.1506429 13, 25, 26, 37, 68, 119

[87] Darrel Hankerson, Alfred J. Menezes, and Scott Vanstone. *Guide to Elliptic Curve Cryptography*. Springer Science & Business Media, 2006. DOI: 10.1007/0-387-23483-7_131 18

[88] Robert Harper, Furio Honsell, and Gordon Plotkin. A Framework for Defining Logics. *Journal of the ACM*, 40(1):143–184, 1993. DOI: 10.1145/138027.138060 108

[89] Chris Hawblitzel, Jon Howell, Jacob R. Lorch, Arjun Narayan, Bryan Parno, Danfeng Zhang, and Brian Zill. Ironclad Apps: End-to-End Security via Automated Full-System Verification. In *Proceedings of the USENIX Symposium on Operating Systems Design and Implementation*, pages 165–181. USENIX Association, 2014. 104, 110

[90] Scott Hazelhurst and Carl-Johan H. Seger. Symbolic Trajectory Evaluation. In *Formal Hardware Verification*, pages 3–78. Springer, 1997. DOI: 10.1007/3-540-63475-4_1 109

[91] Zecheng He and Ruby B. Lee. How Secure is Your Cache Against Side-Channel Attacks? In *Proceedings of the International Symposium on Microarchitecture*, pages 341–353. ACM, 2017. DOI: 10.1145/3123939.3124546 100

[92] Clemens Helfmeier, Christian Boit, and Uwe Kerst. On Charge Sensors for FIB Attack Detection. In *Proceedings of the International Symposium on Hardware-Oriented Security and Trust*, pages 128–133. IEEE, 2012. DOI: 10.1109/hst.2012.6224332 13

[93] Matt Henricksen, Wun She Yap, Chee Hoo Yian, Shinsaku Kiyomoto, and Toshiaki Tanaka. Side-Channel Analysis of the K2 Stream Cipher. In *Information Security and Privacy*, pages 53–73. Springer, 2010. DOI: 10.1007/978-3-642-14081-5_4 91

[94] Matthew Hicks, Cynthia Sturton, Samuel T. King, and Jonathan M. Smith. Specs: A Lightweight Runtime Mechanism for Protecting Software from Security-Critical Processor Bugs. In *ACM SIGARCH Computer Architecture News*, volume 43, pages 517–529. ACM, 2015. DOI: 10.1145/2786763.2694366 29

[95] Matthew Hoekstra, Reshma Lal, Pradeep Pappachan, Vinay Phegade, and Juan Del Cuvillo. Using Innovative Instructions to Create Trustworthy Software Solutions. In *Proceedings of the Workshop on Hardware Support for Security and Privacy*, 2013. DOI: 10.1145/2487726.2488370 39, 48

[96] Weidong Shi Hsien-Hsin, S. Lee Hsien-hsin, Chenghuai Lu, and Mrinmoy Ghosh. Towards the Issues in Architectural Support for Protection of Software Execution. In *Proceedings of the Workshop on Architectureal Support for Security and Anti-Virus*, 2004. DOI: 10.1145/1055626.1055629 71

[97] Wei-Ming Hu. Reducing Timing Channels with Fuzzy Time. In *Proceedings of the IEEE Computer Society Symposium on Research in Security and Privacy*, pages 8–20. IEEE, 1991. DOI: 10.3233/jcs-1992-13-404 101

[98] Wei-Ming Hu. Lattice Scheduling and Covert Channels. In *Proceedings of the IEEE Computer Society Symposium on Research in Security and Privacy*, pages 52–61. IEEE, 1992. DOI: 10.1109/risp.1992.213271 101

[99] Gérard Huet, Gilles Kahn, and Christine Paulin-Mohring. The Coq Proof Assistant: A Tutorial. https://hal.inria.fr/inria-00069918. 108

[100] Casen Hunger, Mikhail Kazdagli, Ankit Rawat, Alex Dimakis, Sriram Vishwanath, and Mohit Tiwari. Understanding Contention-Based Channels and Using Them for Defense. In *Proceedings of the International Symposium of High Performance Computer Architecture*, pages 639–650. IEEE, 2015. DOI: 10.1109/hpca.2015.7056069 96

[101] CryptoCards, IBM Systems cryptographic HSMs. https://www-03.ibm.com/security/cryptocards/. 41

[102] Intel QuickPath Interconnect. https://www.intel.com/content/www/us/en/io/quickpath-technology/quickpath-technology-general.html. 94

[103] Intel 64 and IA-32 Architectures Developer's Manual: Vol. 3B, 2016. http://www.intel.com/design/processor/manuals/253669.pdf. 32

[104] Intel. Intel Software Guard Extensions Developer Guide, 2016. https://download.01.org/intel-sgx/linux-1.7/docs/Intel_SGX_Developer_Guide.pdf. 12, 48, 106

[105] Intel Software Guard Extensions Commercial Licensing FAQ, 2016. https://software.intel.com/en-us/articles/intel-software-guard-extensions-product-licensing-faq. 63

[106] Mohammad A. Islam, Shaolei Ren, and Adam Wierman. Exploiting a Thermal Side Channel for Power Attacks in Multi-Tenant Data Centers. In *Proceedings of the Conference on Computer and Communications Security*, pages 1079–1094. ACM, 2017. DOI: 10.1145/3133956.3133994 10

[107] Yeongjin Jang, Jaehyuk Lee, Sangho Lee, and Taesoo Kim. SGX-Bomb: Locking Down the Processor via Rowhammer Attack. In *Proceedings of the Workshop on System Software for Trusted Execution*, page 5. ACM, 2017. DOI: 10.1145/3152701.3152709 51

[108] Seongwook Jin, Jeongseob Ahn, Sanghoon Cha, and Jaehyuk Huh. Architectural Support for Secure Virtualization Under a Vulnerable Hypervisor. In *Proceedings of the International Symposium on Microarchitecture*, pages 272–283. ACM, 2011. DOI: 10.1145/2155620.2155652 103

[109] Yier Jin, Nathan Kupp, and Yiorgos Makris. Experiences in Hardware Trojan Design and Implementation. In *Proceedings of the International Workshop on Hardware-Oriented Security and Trust*, pages 50–57. IEEE, 2009. DOI: 10.1109/hst.2009.5224971 40, 82

[110] Naghmeh Karimi, Arun Karthik Kanuparthi, Xueyang Wang, Ozgur Sinanoglu, and Ramesh Karri. MAGIC: Malicious Aging in Circuits/Cores. *ACM Transactions on Architecture and Code Optimization*, 12(1), 2015. DOI: 10.1145/2724718 9

[111] Matt Kaufmann and J. Strother Moore. An ACL2 tutorial. In *Theorem Proving in Higher Order Logics*, pages 17–21. Springer, 2008. DOI: 10.1007/978-3-540-71067-7_4 108

[112] Eric Keller, Jakub Szefer, Jennifer Rexford, and Ruby B. Lee. NoHype: Virtualized Cloud Infrastructure Without the Virtualization. In *Proceedings of the International Symposium on Computer Architecture*, pages 350–361. ACM, 2010. DOI: 10.1145/1816038.1816010 38, 49

[113] Georgios Keramidas, Alexandros Antonopoulos, Dimitrios N. Serpanos, and Stefanos Kaxiras. Non Deterministic Caches: A Simple and Effective Defense Against Side Channel Attacks. *Design Automation for Embedded Systems*, 12(3):221–230, 2008. DOI: 10.1007/s10617-008-9018-y 100

[114] Christoph Kern and Mark R. Greenstreet. Formal Verification in Hardware Design: a Survey. *ACM Transactions on Design Automation of Electronic Systems*, 4(2):123–193, 1999. DOI: 10.1145/307988.307989 104

[115] Richard E. Kessler and Mark D. Hill. Page Placement Algorithms for Large Real-Indexed Caches. *ACM Transactions on Computer Systems*, 10(4):338–359, 1992. DOI: 10.1145/138873.138876 100

[116] Taeho Kgil, Laura Falk, and Trevor Mudge. ChipLock: Support for Secure Microarchitectures. *ACM SIGARCH Computer Architecture News*, 33(1):134–143, 2005. DOI: 10.1145/1055626.1055644 103

[117] Yoongu Kim, Ross Daly, Jeremie Kim, Chris Fallin, Ji Hye Lee, Donghyuk Lee, Chris Wilkerson, Konrad Lai, and Onur Mutlu. Flipping Bits in Memory Without Accessing Them: An Experimental Study of DRAM Disturbance Errors. In *ACM SIGARCH Computer Architecture News*, pages 361–372, 2014. DOI: 10.1145/2678373.2665726 9, 25, 26, 119

[118] Gerwin Klein, Kevin Elphinstone, Gernot Heiser, June Andronick, David Cock, Philip Derrin, Dhammika Elkaduwe, Kai Engelhardt, Rafal Kolanski, Michael Norrish, et al. seL4: Formal Verification of an OS Kernel. In *Proceedings of the Symposium on Operating Systems Principles*, pages 207–220. ACM, 2009. DOI: 10.1145/1629575.1629596 104, 110

[119] Ünal Kocabaş, Andreas Peter, Stefan Katzenbeisser, and Ahmad-Reza Sadeghi. *Converse PUF-Based Authentication*. Springer, 2012. DOI: 10.1007/978-3-642-30921-2_9 23

[120] Paul Kocher, Daniel Genkin, Daniel Gruss, Werner Haas, Mike Hamburg, Moritz Lipp, Stefan Mangard, Thomas Prescher, Michael Schwarz, and Yuval Yarom. Spectre Attacks: Exploiting Speculative Execution, January 2018. http://arxiv.org/abs/1801.01203. 2, 26, 28, 29, 87, 103, 120

[121] Paul Kocher, Joshua Jaffe, and Benjamin Jun. Differential Power Analysis. In *Advances in Cryptology – CRYPTO-99*, pages 388–397. Springer, 1999. DOI: 10.1007/3-540-48405-1_25 10

[122] Paul C. Kocher. Timing Attacks on Implementations of Diffie-Hellman, RSA, DSS, and Other Systems. In *Advances in Cryptology – CRYPTO-96*, pages 104–113. Springer, 1996. DOI: 10.1007/3-540-68697-5_9 10

[123] Florian Kohnhäuser, André Schaller, and Stefan Katzenbeisser. PUF-Based Software Protection for Low-End Embedded Devices. In *Trust and Trustworthy Computing*, pages 3–21. Springer, 2015. DOI: 10.1007/978-3-319-22846-4_1 23, 61

[124] Jingfei Kong, Onur Aciiçmez, Jean-Pierre Seifert, and Huiyang Zhou. Hardware-Software Integrated Approaches to Defend Against Software Cache-Based Side Channel Attacks. In *Proceedings of the International Symposium on High Performance Computer Architecture*, pages 393–404. IEEE, 2009. DOI: 10.1109/hpca.2009.4798277 98

[125] Joonho Kong, Farinaz Koushanfar, Praveen K. Pendyala, Ahmad-Reza Sadeghi, and Christian Wachsmann. PUFatt: Embedded Platform Attestation Based on Novel Processor-Based PUFs. In *Proceedings of the Design Automation Conference*, 2014. DOI: 10.1145/2593069.2593192 23

[126] Butler W. Lampson. A Note on the Confinement Problem. *Communications of the ACM*, 16(10):613–615, 1973. DOI: 10.1145/362375.362389 87

[127] Gregor Leander, Erik Zenner, and Philip Hawkes. Cache Timing Analysis of LFSR-Based Stream Ciphers. In *Cryptography and Coding*, pages 433–445. Springer, 2009. DOI: 10.1007/978-3-642-10868-6_26 91

[128] Ruby B. Lee. Security Basics for Computer Architects. *Synthesis Lectures on Computer Architecture*, 8(4):1–111, 2013. DOI: 10.2200/s00512ed1v01y201305cac025 10, 13, 15, 16, 19

[129] Ruby B. Lee, Peter Kwan, John P. McGregor, Jeffrey Dwoskin, and Zhenghong Wang. Architecture for Protecting Critical Secrets in Microprocessors. In *Proceedings of the International Symposium on Computer Architecture*, pages 2–13. IEEE, 2005. DOI: 10.1145/1080695.1069971 36, 38, 47, 48, 100, 103

[130] Sangho Lee, Youngsok Kim, Jangwoo Kim, and Jong Kim. Stealing Webpages Rendered on Your Browser by Exploiting GPU Vulnerabilities. In *IEEE Symposium on Security and Privacy*, pages 19–33. IEEE, 2014. DOI: 10.1109/sp.2014.9 29

[131] K. Rustan and M. Leino. Dafny: an Automatic Program Verifier for Functional Correctness. In *Logic for Programming, Artificial Intelligence, and Reasoning*, pages 348–370. Springer, 2010. DOI: 10.1007/978-3-642-17511-4_20 107, 110

[132] Xavier Leroy. The CompCert C verified Compiler. *Documentation and User's Manual. INRIA Paris-Rocquencourt*, 2012. 107

[133] Xun Li, Vineeth Kashyap, Jason K. Oberg, Mohit Tiwari, Vasanth Ram Rajarathinam, Ryan Kastner, Timothy Sherwood, Ben Hardekopf, and Frederic T. Chong. Sapper: A Language for Hardware-Level Security Policy Enforcement. In *ACM SIGARCH Computer Architecture News*, volume 42, pages 97–112. ACM, 2014. DOI: 10.1145/2541940.2541947 104, 110

[134] Xun Li, Mohit Tiwari, Jason K. Oberg, Vineeth Kashyap, Frederic T. Chong, Timothy Sherwood, and Ben Hardekopf. Caisson: a Hardware Description Language for Secure Information Flow. In *ACM SIGPLAN Notices*, volume 46, pages 109–120. ACM, 2011. DOI: 10.1145/2345156.1993512 104, 110

[135] Libreboot FAQ. https://libreboot.org/faq.html. 37

[136] David Lie, Chandramohan Thekkath, Mark Mitchell, Patrick Lincoln, Dan Boneh, John Mitchell, and Mark Horowitz. Architectural Support for Copy and Tamper Resistant Software. *ACM SIGPLAN Notices*, 35(11):168–177, 2000. DOI: 10.21236/ada419599 36, 38, 47, 103, 104, 109, 110

[137] Chae Hoon Lim and Tymur Korkishko. mCrypton – A Lightweight Block Cipher for Security of Low-Cost RFID Tags and Sensors. In *Proceedings of the International Workshop on Information Security Applications*, volume 3786, pages 243–258. Springer, 2005. DOI: 10.1007/11604938_19 82

[138] Moritz Lipp, Michael Schwarz, Daniel Gruss, Thomas Prescher, Werner Haas, Stefan Mangard, Paul Kocher, Daniel Genkin, Yuval Yarom, and Mike Hamburg. Meltdown, January 2018. https://arxiv.org/abs/1801.01207. 2, 26, 27, 28, 103, 120

[139] Roger Lipsett, Carl F. Schaefer, and Cary Ussery. *VHDL: Hardware Description and Design*. Springer Science & Business Media, 2012. DOI: 10.1007/978-1-4613-1631-2 107

[140] Fangfei Liu, Qian Ge, Yuval Yarom, Frank Mckeen, Carlos Rozas, Gernot Heiser, and Ruby B. Lee. Catalyst: Defeating Last-Level Cache Side Channel Attacks in Cloud Computing. In *Proceedings of the International Symposium on High Performance Computer Architecture*, pages 406–418. IEEE, 2016. DOI: 10.1109/hpca.2016.7446082 49, 100

[141] Fangfei Liu and Ruby B. Lee. Random Fill Cache Architecture. In *Proceedings of the International Symposium on Microarchitecture*, pages 203–215. IEEE, 2014. DOI: 10.1109/micro.2014.28 100

[142] Fangfei Liu, Hao Wu, Kenneth Mai, and Ruby B Lee. Newcache: Secure Cache Architecture Thwarting Cache Side-Channel Attacks. *IEEE Micro*, 36(5):8–16, 2016. DOI: 10.1109/mm.2016.85 100

[143] Roel Maes and Ingrid Verbauwhede. Physically Unclonable Functions: A Study on the State of the Art and Future Research Directions. In *Towards Hardware-Intrinsic Security*, pages 3–37. Springer, 2010. DOI: 10.1007/978-3-642-14452-3_1 23

[144] Abhranil Maiti, Raghunandan Nagesh, Anand Reddy, and Patrick Schaumont. Physical Unclonable Function and True Random Number Generator: a Compact and Scalable Implementation. In *Proceedings of the Great Lakes Symposium on VLSI*, pages 425–428. ACM, 2009. DOI: 10.1145/1531542.1531639 23

[145] Robert Martin, John Demme, and Simha Sethumadhavan. TimeWarp: Rethinking Timekeeping and Performance Monitoring Mechanisms to Mitigate Side-Channel Attacks. In *ACM SIGARCH Computer Architecture News*, volume 40, pages 118–129. IEEE Computer Society, 2012. DOI: 10.1145/2366231.2337173 100

[146] DS5002FP Secure Microprocessor Chip. https://www.maximintegrated.com/en/p roducts/digital/microcontrollers/DS5002FP.html. 39

[147] Michael McCoyd, Robert Bellarmine Krug, Deepak Goel, Mike Dahlin, and William Young. Building a Hypervisor on a Formally Verifiable Protection Layer. In *Proceedings of the Hawaii International Conference on System Sciences*, pages 5069–5078. IEEE, 2013. DOI: 10.1109/hicss.2013.121 104

[148] David McGrew and John Viega. The Galois/Counter Mode of Operation (GCM). *Submission to NIST Modes of Operation Process*, 20, 2004. 16, 77

[149] Frank McKeen, Ilya Alexandrovich, Alex Berenzon, Carlos V Rozas, Hisham Shafi, Vedvyas Shanbhogue, and Uday R Savagaonkar. Innovative Instructions and Software Model for Isolated Execution. In *Proceedings of the International Workshop on Hardware and Architectural Support for Security and Privacy*. ACM, 2013. DOI: 10.1145/2487726.2488368 39, 47, 48, 103

[150] Ralph Merkle. Secrecy, Authentication, and Public Key Systems. *Ph. D. Thesis, Stanford University*, 1979. 20, 71

[151] David L. Mills. Internet Time Synchronization: the Network Time Protocol. *IEEE Transactions on Communications*, 39(10):1482–1493, 1991. DOI: 10.17487/rfc1129 100

[152] CVE-2018-3665 Detail, 2018. `https://cve.mitre.org/cgi-bin/cvename.cgi?name=CVE-2018-3665`. 46

[153] Thomas Moscibroda and Onur Mutlu. Memory Performance Attacks: Denial of Memory Service in Multi-Core Systems. In *Proceedings of the USENIX Security Symposium.* USENIX Association, 2007. 9

[154] Michael Neve and Jean-Pierre Seifert. Advances on Access-Driven Cache Attacks on AES. In *Selected Areas in Cryptography*, pages 147–162. Springer, 2007. DOI: 10.1007/978-3-540-74462-7_11 91

[155] Michael Neve, Jean-Pierre Seifert, and Zhenghong Wang. A Refined Look at Bernstein's AES Side-Channel Analysis. In *Proceedings of the Symposium on Information, Computer and Communications Security*, pages 369–369. ACM, 2006. DOI: 10.1145/1128817.1128887 91

[156] Khang T. Nguyen. Introduction to Cache Allocation Technology in the Intel Xeon Processor E5 v4 Family, 2016. `https://software.intel.com/en-us/articles/introduction-to-cache-allocation-technology`. 100

[157] Rishiyur Nikhil. Bluespec System Verilog: Efficient, Correct RTL from High Level Specifications. In *Proceedings of the International Conference on Formal Methods and Models for Co-Design*, pages 69–70. IEEE, 2004. DOI: 10.1109/memcod.2004.1459818 107

[158] Tobias Nipkow, Lawrence C. Paulson, and Markus Wenzel. *Isabelle/HOL: a Proof Assistant for Higher-Order Logic*, volume 2283. Springer Science & Business Media, 2002. DOI: 10.1007/3-540-45949-9. 108

[159] DoD 5200.28-STD, Department of Defense Trusted Computer System Evaluation Criteria, 1983. `http://csrc.nist.gov/publications/history/dod85.pdf`. 97

[160] Damian L. Osisek, Kathryn M. Jackson, and Peter H. Gum. ESA/390 Interpretive-Execution Architecture, Foundation for VM/ESA. *IBM Systems Journal*, 30(1):34–51, 1991. DOI: 10.1147/sj.301.0034 31

[161] Dag Arne Osvik, Adi Shamir, and Eran Tromer. Cache Attacks and Countermeasures: the Case of AES. In *Topics in Cryptology*, pages 1–20. Springer, 2006. DOI: 10.1007/11605805_1 10, 101

[162] Sam Owre, John M. Rushby, and Natarajan Shankar. PVS: a Prototype Verification System. In *Automated Deduction – CADE-11*, pages 748–752. Springer, 1992. DOI: 10.1007/3-540-55602-8_217 108

[163] Dan Page. Partitioned Cache Architecture as a Side-Channel Defence Mechanism. *IACR Cryptology ePrint Archive*, 2005:280, 2005. 98

[164] David A. Patterson and John L. Hennessy. *Computer Organization and Design RISC-V Edition: The Hardware Software Interface*. Morgan Kaufmann, 2017. 31

[165] Colin Percival. Cache Missing for Fun and Profit, 2005. 10, 91, 96

[166] Nicole Perlroth, Jeff Larson, and Scott Shane. NSA Able to Foil Basic Safeguards of Privacy on Web. *The New York Times*, 5, 2013. 19

[167] Fabien Petitcolas. Kerckhoffs' principles from "La cryptographie militaire". http://pe titcolas.net/kerckhoffs/index.html. 7, 40

[168] Nick L. Petroni Jr., Timothy Fraser, Jesus Molina, and William A. Arbaugh. Copilot – a Coprocessor-Based Kernel Runtime Integrity Monitor. In *Proceedings of the USENIX Security Symposium*, pages 179–194. USENIX Association, 2004. 60

[169] Disabling Intel ME 11 via Undocumented Mode, 2017. http://blog.ptsecurity.co m/2017/08/disabling-intel-me.html. 37

[170] Vaughan Pratt. Anatomy of the Pentium Bug. In *Colloquium on Trees in Algebra and Programming*, pages 97–107. Springer, 1995. DOI: 10.1007/3-540-59293-8_189 25

[171] Adam Procter, William L Harrison, Ian Graves, Michela Becchi, and Gerard All-wein. Semantics Driven Hardware Design, Implementation, and Verification with ReWire. In *ACM SIGPLAN Notices*, volume 50, page 13. ACM, 2015. DOI: 10.1145/2670529.2754970 110

[172] Chester Rebeiro, Debdeep Mukhopadhyay, Junko Takahashi, and Toshinori Fukunaga. Cache Timing Attacks on Clefia. In *Progress in Cryptology–INDOCRYPT 2009*, pages 104–118. Springer, 2009. DOI: 10.1007/978-3-642-10628-6_7 91

[173] Vincent Rijmen and Joan Daemen. Advanced Encryption Standard. *Federal Information Processing Standards Publications, National Institute of Standards and Technology*, pages 19–22, 2001. 17

[174] Thomas Ristenpart, Eran Tromer, Hovav Shacham, and Stefan Savage. Hey, You, Get Off of My Cloud: Exploring Information Leakage in Third-Party Compute Clouds. In *Proceedings of the Conference on Computer and Communications Security*, pages 199–212. ACM, 2009. DOI: 10.1145/1653662.1653687 10, 96

[175] R. L. Rivest, A. Shamir, and L. Adleman. A Method for Obtaining Digital Signatures and Public-key Cryptosystems. *Communications of the ACM*, 21(2):120–126, February 1978. DOI: 10.21236/ada606588 17, 18

[176] Brian Rogers, Siddhartha Chhabra, Milos Prvulovic, and Yan Solihin. Using Address Independent Seed Encryption and Bonsai Merkle Trees to Make Secure Processors OS-and Performance-Friendly. In *Proceedings of the International Symposium on Microarchitecture*, pages 183–196. IEEE, 2007. DOI: 10.1109/micro.2007.4408255 72

[177] Brian Rogers, Milos Prvulovic, and Yan Solihin. Efficient Data Protection for Distributed Shared Memory Multiprocessors. In *Proceedings of the International Conference of Parallel Architectures and Compilation Techniques*, pages 84–94. IEEE, 2006. DOI: 10.1145/1152154.1152170 71, 79

[178] Masoud Rostami, Farinaz Koushanfar, Jeyavijayan Rajendran, and Ramesh Karri. Hardware Security: Threat Models and Metrics. In *Proceedings of the International Conference on Computer-Aided Design*, pages 819–823. IEEE Press, 2013. DOI: 10.1109/iccad.2013.6691207 6, 13, 40

[179] Alexander Rostovtsev and Anton Stolbunov. Public-Key Cryptosystem Based on Isogenies. Cryptology ePrint Archive, Report 2006/145, 2006. http://eprint.iacr.org/2006/145. 18

[180] Jarrod A. Roy, Farinaz Koushanfar, and Igor L. Markov. EPIC: Ending Piracy of Integrated Circuits. In *Proceedings of the Conference on Design, Automation and Test in Europe*, pages 1069–1074. ACM, 2008. DOI: 10.1109/date.2008.4484823 40

[181] Xiaoyu Ruan. *Platform Embedded Security Technology Revealed: Safeguarding the Future of Computing with Intel Embedded Security and Management Engine*. Apress, 2014. DOI: 10.1007/978-1-4302-6572-6 33

[182] Ulrich Ruhrmair, J. L. Martinez-Hurtado, Xiaolin Xu, Christian Kraeh, Christian Hilgers, Dima Kononchuk, Jonathan J. Finley, and Wayne P. Burleson. Virtual Proofs of Reality and Their Physical Implementation. In *IEEE Symposium on Security and Privacy*, pages 70–85. IEEE, 2015. DOI: 10.1109/sp.2015.12 23

[183] Joanna Rutkowska. Intel x86 Considered Harmful, 2015. https://blog.invisiblethings.org/papers/2015/x86_harmful.pdf. 33

[184] A. Sabelfeld and A.C. Myers. Language-Based Information-Flow Security. *IEEE Journal on Selected Areas in Communications*, 21(1):5–19, January 2003. DOI: 10.1109/jsac.2002.806121 110

[185] Daniel Sanchez and Christos Kozyrakis. The ZCache: Decoupling Ways and Associativity. In *Proceedings of the International Symposium on Microarchitecture*, pages 187–198. IEEE, 2010. DOI: 10.1109/micro.2010.20 92, 99

[186] André Schaller, Tolga Arul, Vincent van der Leest, and Stefan Katzenbeisser. Lightweight Anti-counterfeiting Solution for Low-End Commodity Hardware Using Inherent PUFs. In *Trust and Trustworthy Computing*, pages 83–100. Springer, 2014. DOI: 10.1007/978-3-319-08593-7_6 23, 61

[187] Ryan A. Scheel and Akhilesh Tyagi. Characterizing Composite User-Device Touchscreen Physical Unclonable Functions (PUFs) for Mobile Device Authentication. In *Proceedings of the International Workshop on Trustworthy Embedded Devices*, pages 3–13. ACM, 2015. DOI: 10.1145/2808414.2808418 23, 61

[188] Michael D. Schroeder and Jerome H. Saltzer. A Hardware Architecture for Implementing Protection Rings. *Communications of the ACM*, 15(3):157–170, 1972. DOI: 10.1145/361268.361275 30

[189] Steffen Schulz, Ahmad-Reza Sadeghi, and Christian Wachsmann. Short Paper: Lightweight Remote Attestation Using Physical Functions. In *Proceedings of the Conference on Wireless Network Security*, pages 109–114. ACM, 2011. DOI: 10.1145/1998412.1998432 23

[190] Oliver Schwarz and Mads Dam. Automatic Derivation of Platform Noninterference Properties. In *Proceedings of the International Conference on Software Engineering and Formal Methods*, pages 27–44. Springer, 2016. DOI: 10.1007/978-3-319-41591-8_3 110

[191] Arvind Seshadri, Mark Luk, Ning Qu, and Adrian Perrig. SecVisor: A Tiny Hypervisor to Provide Lifetime Kernel Code Integrity for Commodity OSes. *ACM SIGOPS Operating Systems Review*, 41(6):335–350, 2007. DOI: 10.1145/1323293.1294294 104

[192] Ofer Shacham, Megan Wachs, Andrew Danowitz, Sameh Galal, John Brunhaver, Wajahat Qadeer, Sabarish Sankaranarayanan, Artem Vassiliev, Stephen Richardson, and Mark Horowitz. Avoiding Game Over: Bringing Design to the Next Level. In *Proceedings of the Annual Design Automation Conference*, pages 623–629. ACM, 2012. DOI: 10.1145/2228360.2228472 107

[193] John Paul Shen and Mikko H. Lipasti. *Modern processor design: fundamentals of superscalar processors*. Waveland Press, 2013. 93

[194] Weidong Shi, H. S. Lee, M. Ghosh, Chenghuai Lu, and A. Boldyreva. High Efficiency Counter Mode Security Architecture via Prediction and Precomputation. In *Proceedings of the International Symposium on Computer Architecture*, June 2005. DOI: 10.1109/isca.2005.30 71

[195] Weidong Shi, Hsien-Hsin S Lee, Mrinmoy Ghosh, and Chenghuai Lu. Architectural Support for High Speed Protection of Memory Integrity and Confidentiality in Multiprocessor Systems. In *Proceedings of the International Conference on Parallel Architectures and Compilation Techniques*, pages 123–134. IEEE Computer Society, 2004. DOI: 10.1109/pact.2004.1342547 71, 77

[196] Kanna Shimizu, H. Peter Hofstee, and John S. Liberty. Cell Broadband Engine Processor Vault Security Architecture. *IBM Journal of Research and Development*, 51(5):521–528, 2007. DOI: 10.1147/rd.515.0521 39, 48

[197] Peter W. Shor. Algorithms for Quantum Computation: Discrete Logarithms and Factoring. In *Foundations of Computer Science*, pages 124–134. IEEE, 1994. DOI: 10.1109/sfcs.1994.365700 18

[198] Peter W. Shor. Polynomial-Time Algorithms for Prime Factorization and Discrete Logarithms on a Quantum Computer. *SIAM Review*, 41(2):303–332, 1999. DOI: 10.1137/s0036144598347011 18

[199] Vijay D. Silva, Daniel Kroening, and Georg Weissenbacher. A Survey of Automated Techniques for Formal Software Verification. *IEEE Transactions on Computer-Aided Design of Integrated Circuits and Systems*, 27(7):1165–1178, 2008. DOI: 10.1109/tcad.2008.923410 104

[200] Rohit Sinha, Manuel Costa, Akash Lal, Nuno P. Lopes, Sriram Rajamani, Sanjit A Seshia, and Kapil Vaswani. A Design and Verification Methodology for Secure Isolated Regions. In *Proceedings of the Conference on Programming Language Design and Implementation*, pages 665–681. ACM, 2016. DOI: 10.1145/2908080.2908113 110

[201] Rohit Sinha, Sriram Rajamani, Sanjit Seshia, and Kapil Vaswani. Moat: Verifying Confidentiality of Enclave Programs. In *Proceedings of the Conference on Computer and Communications Security*, pages 1169–1184. ACM, 2015. DOI: 10.1145/2810103.2813608 104, 105, 110

[202] Secure Hash Standard. FIPS PUB 180-2. *National Institute of Standards and Technology*, 2002. DOI: 10.1080/01611194.2012.687431. 21

[203] Emil Stefanov, Marten Van Dijk, Elaine Shi, Christopher Fletcher, Ling Ren, Xiangyao Yu, and Srinivas Devadas. Path ORAM: an Extremely Simple Oblivious RAM Protocol. In *Proceedings of the Conference on Computer and Communications Security*, pages 299–310. ACM, 2013. DOI: 10.1145/2508859.2516660 73

[204] G. Edward Suh, Dwaine Clarke, Blaise Gassend, Marten van Dijk, and Srinivas Devadas. Efficient Memory Integrity Verification and Encryption for Secure Processors. In

Proceedings of the International Symposium on Microarchitecture. IEEE Computer Society, 2003. DOI: 10.1109/micro.2003.1253207 71

[205] G. Edward Suh, Dwaine Clarke, Blaise Gassend, Marten Van Dijk, and Srinivas Devadas. AEGIS: Architecture for Tamper-Evident and Tamper-Resistant Processing. In *Proceedings of the International Conference on Supercomputing*, pages 160–171. ACM, 2003. DOI: 10.1145/782837.782838 36, 38, 47, 71, 103

[206] G. Edward Suh and Srinivas Devadas. Physical Unclonable Functions for Device Authentication and Secret Key Generation. In *Proceedings of the Design Automation Conference*, pages 9–14, 2007. DOI: 10.1109/dac.2007.375043 23

[207] Jakub Szefer. Survey of Microarchitectural Side and Covert Channels, Attacks, and Defenses. Cryptology ePrint Archive, Report 2016/479, 2016. http://eprint.iacr.org/2016/479. 85

[208] Jakub Szefer, Eric Keller, Ruby B. Lee, and Jennifer Rexford. Eliminating the Hypervisor Attack Surface for a More Secure Cloud. In *Proceedings of the Conference on Computer and Communications Security*, pages 401–412. ACM, 2011. DOI: 10.1145/2046707.2046754 38, 49

[209] Jakub Szefer and Ruby B. Lee. Architectural Support for Hypervisor-Secure Virtualization. In *ACM SIGPLAN Notices*, volume 47, pages 437–450. ACM, 2012. DOI: 10.1145/2189750.2151022 36, 38, 45, 49, 103, 105

[210] Adrian Tang, Simha Sethumadhavan, and Salvatore Stolfo. CLKSCREW: Exposing the Perils of Security-Oblivious Energy Management. In *Proceedings of the USENIX Security Symposium*. USENIX Association, 2017. 29

[211] George Taylor, Peter Davies, and Michael Farmwald. The TLB Slice – a Low-Cost High-Speed Address Translation Mechanism. In *Proceedings of the International Symposium on Computer Architecture*, pages 355–363. IEEE, 1990. DOI: 10.1109/isca.1990.134546 100

[212] Mohammad Tehranipoor and Farinaz Koushanfar. A Survey of Hardware Trojan Taxonomy and Detection. *IEEE Design & Test of Computers*, 27(1), 2010. DOI: 10.1109/mdt.2009.159 40

[213] Mohammad Tehranipoor and Cliff Wang. *Introduction to Hardware Security and Trust*. Springer Science & Business Media, 2011. DOI: 10.1007/978-1-4419-8080-9 6, 13

[214] Alexander Tereshkin and Rafal Wojtczuk. Introducing Ring -3 Rootkits, 2009. http://invisiblethingslab.com/itl/Resources.html. 50

[215] Donald Thomas and Philip Moorby. *The Verilog Hardware Description Language*. Springer Science & Business Media, 2008. DOI: 10.1007/978-1-4757-2365-6 107

[216] Ken Thompson. Reflections on Trusting Trust. *Communications of the ACM*, 27(8):761–763, 1984. DOI: 10.1145/1283920.1283940 107

[217] Kris Tiri, Onur Acıiçmez, Michael Neve, and Flemming Andersen. An Analytical Model for Time-Driven Cache Attacks. In *Proceedings of the Fast Software Encryption*, pages 399–413. Springer, 2007. DOI: 10.1007/978-3-540-74619-5_25 91

[218] Mohit Tiwari, Xun Li, Hassan M. G. Wassel, Frederic T. Chong, and Timothy Sherwood. Execution Leases: A Hardware-Supported Mechanism for Enforcing Strong Non-Interference. In *Proceedings of the International Symposium on Microarchitecture*, pages 493–504. ACM, 2009. DOI: 10.1145/1669112.1669174 99

[219] Trusted Computing Group Trusted Platform Module Main Specification Version 1.2, Revision 94. http://www.trustedcomputinggroup.org/resources/tpm_main_spe cification. 56

[220] Eran Tromer, Dag Arne Osvik, and Adi Shamir. Efficient Cache Attacks on AES, and Countermeasures. *Journal of Cryptology*, 23(1):37–71, 2010. DOI: 10.1007/s00145-009-9049-y 91

[221] ARM Trustzone. TrustZone Information Page, 2016. http://www.arm.com/produc ts/processors/technologies/trustzone/. 103

[222] Yukiyasu Tsunoo, Teruo Saito, Tomoyasu Suzaki, Maki Shigeri, and Hiroshi Miyauchi. Cryptanalysis of DES Implemented on Computers with Cache. In *Cryptographic Hardware and Embedded Systems*, pages 62–76. Springer, 2003. DOI: 10.1007/978-3-540-45238-6_6 91

[223] Yukiyasu Tsunoo, Teruo Saito, Tomoyasu Suzaki, Maki Shigeri, and Hiroshi Miyauchi. Cryptanalysis of DES Implemented on Computers with Cache. In *Proceedings of the International Workshop on Cryptographic Hardware and Embedded Systems*, pages 62–76. Springer, 2003. DOI: 10.1007/978-3-540-45238-6_6 91

[224] Pim Tuyls and Lejla Batina. RFID-Tags for Anti-Counterfeiting. In *Topics in Cryptology*, pages 115–131. Springer, 2006. DOI: 10.1007/11605805_8 23

[225] Pim Tuyls, Geert-Jan Schrijen, Frans Willems, Tanya Ignatenko, and B. Skoric. Secure Key Storage with PUFs. *Security with Noisy Data – On Private Biometrics, Secure Key Storage and Anti-Counterfeiting*, pages 269–292, 2007. DOI: 10.1007/978-1-84628-984-2_16 23

[226] Pim Tuyls and Boris Škorić. Secret Key Generation from Classical Physics: Physical Uncloneable Functions. In *AmIware Hardware Technology Drivers of Ambient Intelligence*, pages 421–447. Springer, 2006. DOI: 10.1007/1-4020-4198-5_20 23

[227] Rich Uhlig, Gil Neiger, Dion Rodgers, Amy L. Santoni, Fernando C. M. Martins, Andrew V. Anderson, Steven M. Bennett, Alain Kagi, Felix H. Leung, and Larry Smith. Intel Virtualization Technology. *Computer*, 38(5):48–56, 2005. DOI: 10.1109/mc.2005.163 31

[228] Amit Vasudevan, Sagar Chaki, Limin Jia, Jonathan McCune, James Newsome, and Anupam Datta. Design, Implementation and Verification of an Extensible and Modular Hypervisor Framework. In *Proceedings of the Symposium on Security and Privacy*, pages 430–444. IEEE, 2013. DOI: 10.1109/sp.2013.36 104

[229] Yao Wang, Andrew Ferraiuolo, Danfeng Zhang, Andrew C. Myers, and G. Edward Suh. SecDCP: Secure Dynamic Cache Partitioning for Efficient Timing Channel Protection. In *Proceedings of the Design Automation Conference*. IEEE, 2016. DOI: 10.1145/2897937.2898086 100

[230] Yao Wang and G. Edward Suh. Efficient Timing Channel Protection for On-Chip Networks. In *Proceedings of the International Symposium on Networks on Chip*, pages 142–151. IEEE, 2012. DOI: 10.1109/nocs.2012.24 99

[231] Yujie Wang, Pu Chen, Jiang Hu, Guofeng Li, and Jeyavijayan Rajendran. The Cat and Mouse in Split Manufacturing. *IEEE Transactions on Very Large Scale Integration Systems*, 26(5):805–817, 2018. DOI: 10.1145/2897937.2898104 6, 13, 40

[232] Zhenghong Wang and Ruby B. Lee. New Cache Designs for Thwarting Software Cache-based Side Channel Attacks. In *ACM SIGARCH Computer Architecture News*, volume 35, pages 494–505. ACM, 2007. DOI: 10.1145/1273440.1250723 92, 98, 100

[233] Zhenghong Wang and Ruby B. Lee. A Novel Cache Architecture with Enhanced Performance and Security. In *Proceedings of the International Symposium on Microarchitecture*, pages 83–93. IEEE, 2008. DOI: 10.1109/micro.2008.4771781 98, 100

[234] Hassan M. G. Wassel, Ying Gao, Jason K. Oberg, Ted Huffmire, Ryan Kastner, Frederic T. Chong, and Timothy Sherwood. SurfNoC: a Low Latency and Provably Non-Interfering Approach to Secure Networks-on-Chip. *ACM SIGARCH Computer Architecture News*, 41(3):583–594, 2013. DOI: 10.1145/2508148.2485972 99

[235] Matthew M. Wilding, David A. Greve, Raymond J. Richards, and David S. Hardin. Formal Verification of Partition Management for the AAMP7G Microprocessor. In *Design and Verification of Microprocessor Systems for High-Assurance Applications*, pages 175–191. Springer, 2010. DOI: 10.1007/978-1-4419-1539-9_6 110

[236] Johannes Winter.   Experimenting with ARM TrustZone–Or: How I Met Friendly Piece of Trusted Hardware.   In *International Conference on Trust, Security and Privacy in Computing and Communications*, pages 1161–1166. IEEE, 2012. DOI: 10.1109/trustcom.2012.157 39, 48

[237] Rafal Wojtczuk and Joanna Rutkowska. Attacking SMM Memory via Intel CPU Cache Poisoning, 2009. `https://invisiblethingslab.com/resources/misc09/smm_cache_fun.pdf`. 29, 36

[238] Rafal Wojtczuk and Joanna Rutkowska. Following the White Rabbit: Software Attacks Against Intel VT-d, 2011. `https://theinvisiblethings.blogspot.com/2011/05/following-white-rabbit-software-attacks.html`. 29

[239] Jonathan Woodruff, Robert N. M. Watson, David Chisnall, Simon W. Moore, Jonathan Anderson, Brooks Davis, Ben Laurie, Peter G. Neumann, Robert Norton, and Michael Roe. The CHERI Capability Model: Revisiting RISC in an Age of Risk. In *Proceeding of the International Symposium on Computer Architecuture*, pages 457–468. IEEE Press, 2014. DOI: 10.1145/2678373.2665740 34, 38, 103

[240] Weiyi Wu and Bryan Ford. Deterministically Deterring Timing Attacks in Deterland. *arXiv preprint arXiv:1504.07070*, 2015. 101

[241] Zhenyu Wu, Zhang Xu, and Haining Wang. Whispers in the Hyper-space: High-speed Covert Channel Attacks in the Cloud. In *Proceedings of the USENIX Security Symposium*, pages 159–173. USENIX Association, 2012. 96

[242] Wenjie Xiong, Andre Schaller, Nikolaos A. Anagnostopoulos, Muhammad Umair Saleem, Sebastian Gabmeyer, Stefan Katzenbeisser, and Jakub Szefer. Run-Time Accessible DRAM PUFs in Commodity Devices. In *Proceedings of the Conference on Cryptographic Hardware and Embedded Systems*, August 2016. DOI: 10.1007/978-3-662-53140-2_21 61

[243] Yunjing Xu, Michael Bailey, Farnam Jahanian, Kaustubh Joshi, Matti Hiltunen, and Richard Schlichting. An Exploration of L2 Cache Covert Channels in Virtualized Environments. In *Proceedings of the Workshop on Cloud Computing Security Workshop*, pages 29–40. ACM, 2011. DOI: 10.1145/2046660.2046670 96

[244] Chenyu Yan, Daniel Englender, Milos Prvulovic, Brian Rogers, and Yan Solihin. Improving Cost, Performance, and Security of Memory Encryption and Authentication. In *Proceedings of the International Symposium on Computer Architecture*, pages 179–190. IEEE Computer Society, 2006. DOI: 10.1145/1150019.1136502 71

[245] Mengjia Yan, Bhargava Gopireddy, Thomas Shull, and Josep Torrellas. Secure Hierarchy-Aware Cache Replacement Policy (SHARP): Defending Against Cache-Based Side

Channel Attacks. In *Proceedings of the International Symposium on Computer Architecture*, pages 347–360. ACM, 2017. DOI: 10.1145/3079856.3080222 100

[246] Jean Yang and Chris Hawblitzel. Safe to the Last Instruction: Automated Verification of a Type-Safe Operating System. In *ACM Sigplan Notices*, volume 45, pages 99–110. ACM, 2010. DOI: 10.1145/1809028.1806610 104, 110

[247] Jun Yang, Youtao Zhang, and Lan Gao. Fast Secure Processor for Inhibiting Software Piracy and Tampering. In *Proceedings of the International Symposium on Microarchitecture*. IEEE Computer Society, 2003. DOI: 10.1109/micro.2003.1253209 72

[248] Erik Zenner. A Cache Timing Analysis of HC-256. In *Selected Areas in Cryptography*, pages 199–213. Springer, 2009. DOI: 10.1007/978-3-642-04159-4_13 91

[249] Danfeng Zhang, Aslan Askarov, and Andrew C. Myers. Language-Based Control and Mitigation of Timing Channels. *ACM SIGPLAN Notices*, 47(6):99–110, 2012. DOI: 10.1145/2345156.2254078 100

[250] Danfeng Zhang, Yao Wang, G. Edward Suh, and Andrew C. Myers. A Hardware Design Language for Timing-Sensitive Information-Flow Security. In *ACM SIGARCH Computer Architecture News*, volume 43, pages 503–516. ACM, 2015. DOI: 10.1145/2786763.2694372 100, 104, 107, 110

[251] Tianwei Zhang and Ruby B. Lee. New Models of Cache Architectures Characterizing Information Leakage from Cache Side Channels. In *Proceedings of the Annual Computer Security Applications Conference*, pages 96–105. ACM, 2014. DOI: 10.1145/2664243.2664273 104, 109, 110

[252] Yinqian Zhang, Ari Juels, Alina Oprea, and Michael K. Reiter. Homealone: Co-Residency Detection in the Cloud via Side-Channel Analysis. In *Proceedings of the Symposium on Security and Privacy*, pages 313–328. IEEE, 2011. DOI: 10.1109/sp.2011.31 91

[253] Yinqian Zhang, Ari Juels, Michael K. Reiter, and Thomas Ristenpart. Cross-VM Side Channels and Their Use to Extract Private Keys. In *Proceedings of the Conference on Computer and Communications Security*, pages 305–316. ACM, 2012. DOI: 10.1145/2382196.2382230 10, 96

[254] Youtao Zhang, Lan Gao, Jun Yang, Xiangyu Zhang, and Rajiv Gupta. SENSS: Security Enhancement to Symmetric Shared Memory Multiprocessors. In *Proceedings of the International Symposium on High-Performance Computer Architecture*, pages 352–362. IEEE, 2005. DOI: 10.1109/hpca.2005.31 72, 77

Online Resources

BOOK'S WEB PAGE

Online recourses related to this book are available at: `caslab.csl.yale.edu/books`. The web page provides information on book updates, tutorials relating to secure processor architectures and design, and other books or book chapters written by the author.

Author's Biography

JAKUB SZEFER

Jakub Szefer's research interests are at the intersection of computer architecture and hardware security. Jakub's recent projects focus on security verification of processor architectures; hardware (FPGA) implementation of cryptographic algorithms, especially post-quantum cryptographic (PQC) algorithms; Cloud FPGA security; designs of new Physically Unclonable Functions (PUFs); and leveraging physical properties of computer hardware for new cryptographic and security applications. Jakub's research is currently supported through National Science Foundation and industry donations. Jakub is a recipient of a 2017 NSF CAREER award. In the summer of 2013, he became an Assistant Professor of Electrical Engineering at Yale University, where he started the Computer Architecture and Security Laboratory (CAS Lab). Prior to joining Yale, he received Ph.D. and M.A. degrees in Electrical Engineering from Princeton University, where he worked with his advisor, Prof. Ruby B. Lee, on secure processor architectures. He received a B.S. with highest honors in Electrical and Computer Engineering from the University of Illinois at Urbana-Champaign. He can be reached at: jakub.szefer@yale.edu.

Printed in the United States
by Baker & Taylor Publisher Services